LAST TRAIN TO HILVERSUM

LAST TRAIN TO HILVERSUM

A journey in search of the magic of radio

CHARLIE CONNELLY

BLOOMSBURY PUBLISHING
LONDON · OXFORD · NEW YORK · NEW DELHI · SYDNEY

BLOOMSBURY PUBLISHING
Bloomsbury Publishing Plc
50 Bedford Square, London, WC1B 3DP, UK

BLOOMSBURY, BLOOMSBURY PUBLISHING and the Diana logo are trademarks of
Bloomsbury Publishing Plc

First published in Great Britain 2019
This edition published 2019

A catalogue record for this book is available from the British Library

Library of Congress Cataloguing-in-Publication data has been applied for

ISBN: HB: 978-1-4088-8999-2; PB: 978-1-4088-9000-4; eBook: 978-1-4088-8998-5

2 4 6 8 10 9 7 5 3 1

Typeset by Deanta Global Publishing Services, Chennai, India
Printed and bound in Great Britain by CPI Group (UK) Ltd, Croydon CR0 4YY

To find out more about our authors and books visit www.bloomsbury.com and
sign up for our newsletters

As one who, walking in the twilight gloom,

Hears round about him voices as it darkens,
And seeing not the forms from which they come,
Pauses from time to time, and turns and hearkens;

So walking here in twilight, O my friends!
I hear your voices, softened by the distance,
And pause, and turn to listen, as each sends
His words of friendship, comfort, and assistance.

Henry Wadsworth Longfellow

CONTENTS

CONTENTS

1

We Are All Radio People

We are all radio people. I am, you are, the postman is; those two lads over there in the van, that woman with the toddler in the buggy, the old lady next door. All of them, all of us, radio people. According to Radio Joint Audience Research, the independent body that measures radio audiences in Britain, 90 per cent of the adult population listens to the radio every week. Not once a year or a couple of times a month: every week. That's 48.9 million of us, according to RAJAR figures released at the start of 2018, racking up a billion hours a week of radio in our cars, workshops and kitchens, on our transistors and DAB radios, through our phones, laptops and digital televisions. Just about all of us are listening to the radio every week whether we're tuning in to the shipping forecast or an urban grime pirate station. Sixty per cent of us listen at home, a quarter of us listen in the car, the rest of us have the radio on while we're working, all helping to register a frankly astonishing weekly average of 21.3 hours per listener.

These are barely believable statistics – and I'm keen to know who the ten per cent *not* listening to the radio are, the weirdos – especially considering how much more we pick and choose what media we consume and access it at our own convenience these days. Despite the prevalence of podcasts, streaming and the playlists we construct for our mp3 players, radio is still everywhere. It's the oldest broadcasting format in the world (although the kind of radio we listen to today is still less than a hundred years old) yet it's battering through the twenty-first century stronger than ever, adapting to and embracing new technologies while never losing that sense of being the box of voices, the reassuring portal to the world in the corner of the room or the centre of the dashboard.

Let's not forget too that we're not paying a penny for the billion hours of radio we're listening to every week either. Radio, whether it's the BBC or commercial, is completely free. While television rifles your pockets for access to many of its wonders there is no concept of subscription radio or 'pay-per-listen'. Even BBC radio is free: while the corporation's services are funded by the licence fee, you only have to have a licence if you're watching television. Get rid of your telly, stop paying for a licence and nobody's going to come hammering on your door for sticking on Chris Evans in the morning, or pull you over to rat-a-tat-tat on your driver's window for singing along at the wheel to New Order on 6 Music (unless your singing is particularly rotten).

Radio brings the world right into our homes and doesn't ask a penny for it. The news, live football matches, the weather forecast, concerts, documentaries, a new band you've never heard before, every ball of a Test match taking place on the other side of the world, that song you had played at your wedding . . . it's there if you want it and it's all yours for nothing. *Nothing.*

At the same time, podcasts and 'listen again' notwithstanding, radio is probably the last media and entertainment outlet we consume from which we cede control almost completely. We're much choosier about television, becoming less and less likely to allow our viewing to be dictated by the published schedule. In the age of the boxed set and 'series record' buttons the days of parking in front of *Blue Peter* and just watching whatever's on all the way through *Home and Away*, *Top of the Pops*, *Coronation Street* and *Sportsnight* are long gone. Instead we'll flick around the programme guide to see if there's anything on, catch up with the *Marcella* we recorded the night before, or the Saturday night BBC Four Scandinavian drama, then spend at least half an hour scrolling through Netflix trying to find something we want to watch. Yet still we're more likely to put on the radio and succumb to the disembodied voices that fill the room or the car, especially if we're doing something else with the radio on in the background. If it's our local station there'll be news and phone-ins relevant to where we are and we might learn some snippet of local history, what's on at the comedy club in town at the weekend and why someone called Dave thinks Brexit will be brilliant for the minicab firm by the railway station. On Radio 4 we'll hear a politician squirm, listen to a documentary about poetry, and a drama we don't really understand, but we'll leave it on anyway, while on a music station we'll hear a brilliant band we've never heard before, Google them and buy their album from Bandcamp.

Television moors us to one place, requiring a physical engagement as well as a mental one. You have to be in front of the television, it ties us down, but radio sets us free. How often do we just sit down and listen to the radio in the same way we watch the television? Most of us are doing something else – cooking, tidying up, making something, going out for a walk, washing up – allowing the radio to take our minds elsewhere while our bodies are getting on with more mundane stuff for which they don't really need us.

It's a hoary old cliché to say that the pictures are better on the radio, but as with every cliché there's an essential truth at the heart of it. Part of the magic of radio is that we conjure the pictures ourselves. The voices prompt them, we create them. There are radio presenters I've listened to for years and have absolutely no idea what they look like and have no desire to know what they look like either. I have an image in my head, an unfinished sketch, an impression of how I think they look, and that's all I need. I don't need pictures, I don't want them. We're fed so many images in everyday life that I'd rather leave radio unseen, as it was intended.

I'm not a regular listener to *The Archers* but my parents always had the omnibus on every Sunday, so it's been there in the background for almost my entire life. Listening to it recently after a bit of a lull I realised I've planned and laid out Ambridge in my mind almost subconsciously. I can pretty much describe the horse brasses and prints hanging on the walls in the Bull and smell the slight mustiness among the shelves of the shop, yet I bet if I dropped you off in my Ambridge it would be completely unrecognisable from yours. Similarly, if I took a bus from Borchester and alighted at your Ambridge I'd probably think I'd got off at the wrong stop and disrupt an important scene by asking where I was. But we've all heard the same voices and followed the same storylines, for decades in some cases (Eddie Grundy looks a bit like Danny Baker in my Ambridge, by the way, while there's a definite air of Anita Harris about Shula).

The fact that we're conjuring such pictures ourselves creates a unique kind of intimacy with radio. There's a one-to-one relationship between us and the voices and sounds coming out of our radios in a way there isn't with television. With television we feel more part of a crowd or a group of people; we're on the outside looking in at something. With radio it's a different sensation, more of an immersion, a stronger bond somehow, maybe because we've allowed radio to penetrate further into our minds, investing more of ourselves in the experience because we're absorbing sounds and creating images at the same time. It's a collaboration: the radio provides the sound, we provide the pictures.

I'm fortunate enough to have experienced this relationship from both sides. I've been an avid radio listener for my whole life and that experience has been a wholly beneficial one. It's helped me through the regular periods of cripplingly low self-esteem, intense shyness and depression that have dragged me down right back to childhood. At times when I've been, at best, intimidated by the world and, at worst, unable to go out of the front door and participate in it, the radio has been my portal, my vehicle to winch myself back up from darkness to light. The mental health benefits of radio have been, to me, incalculable.

I've also been lucky enough to make radio myself. I've created, written and presented programmes and documentaries on the radio, and co-presented three series alongside a bona fide radio national treasure. I've written a comedy panel show that went out on national radio in Ireland and turned out sketches for radio on both sides of the Irish Sea, conjuring up images and scenarios that induced laughter from complete strangers – a feeling that's hard to beat. I wouldn't call myself a radio professional by any means, but the writing and broadcasting I've done has allowed me to retain a sense of the magic of the medium. I've never lost the thrill of making radio; it's never become a chore or a routine and it's never dimmed the love I have for the voices in the ether.

But for all my enthusiasm for listening to the radio and for making it, it was only recently that I began to think about the vital role it has in our lives, partly because we're on the cusp of a crucial moment in the radio story. Radio is changing because the way we listen to the radio is changing. In 2016 the number of people listening to digital radio at home exceeded the number of listeners on traditional analogue sets for the first time, bringing the day when FM and AM analogue radio is switched off forever a step closer. In Britain three criteria have to be met for this to happen: digital radio's geographical coverage needs to match that of FM; local radio must be available to 90 per cent of the population; and listeners on digital need to comprise at least 50 per cent of the total listening population. Once all this is in place a two-year countdown will begin to the Great Switch-Off. The spring of 2018 saw digital radio register 50.9 per cent of total radio listening to tick off one of those three requirements, but the BBC's director of radio and music, Bob Shennan, responded by advocating for a choice of formats and suggesting the discussion of switching off FM be postponed until a 'review of the landscape in a few years' time'. It's a vague punting of the issue up the street but the switch-off is coming, for sure, it's just a question of when.

It's not Luddite to feel sadness at the inevitable move towards an entirely digital global radio because the range and sound quality available to everyone is already astonishing. From local stations to event-specific stations to specialist music stations, just about everyone is catered for as long as they have access via digital television, radio or the internet. The success of digital-only stations in this country such as the BBC's 6 Music, Asian Network and Radio 4 Extra means that you can hear music of which John Peel would largely have approved, British-Indian dramas and re-runs of classic radio comedy at any time of the day or night. BBC radio shows are available online usually for at least a month after broadcast, if not permanently, which can only be a good thing: with analogue radio once something's broadcast it's gone, you can't rewind or listen again.

Digital is not perfect, however. When the pips go at the top of the hour, for example, and I have the radio running through the internet, television and a regular transistor radio in different parts of the flat, they are all at different times. If I stand in the right place I can hear 18 pips instead of the standard six, and have no idea what the correct time is. Which ones are accurate? My money's on analogue. Internet radio can buffer, broadband signals and cable connections can be knocked out, not to mention full-on power cuts. Batteries and even clockwork make pre-digital radios more reliable in real terms.

While excited about the possibilities of a digital radio future, I can't help feeling a slight tinge of sadness for a passing age. The pop and crackle is on its way out. FM will be switched off, AM will go too, and long wave's days are surely numbered – not least because the main transmitter at Droitwich in Worcestershire is by all accounts on its last legs. One day there will be no dial to turn, no in-between spaces on the waveband for washes of static, ethereal beeps and faint, distant voices. The mystery will be gone: we'll always know exactly what it is we're listening to, whether it's via scrolling LCD on our digital radios, the panel at the bottom of our TV screen or because we've gone in search of a particular streaming station online.

The essential radio truths will still prevail, of course. The entrepreneurs featured on *Desert Island Discs* will continue to have startlingly dull taste in music (can 'The Way It Is' by Bruce Hornsby and the Range really be that inspiring to so many business leaders?) and people calling radio shows will still insist on saying 'hello to anyone who knows me', but the specialisation that comes with the digitalisation of radio means we may lose much of the serendipity of radio, of catching programmes on subjects we never considered before, or hearing a genre of music for the first time that sparks a whole new enthusiasm.

That's not to say digital can't develop the way it disseminates itself, of course. One of the best developments in digital radio recently has been the appearance of websites like Global Breakfast Radio and the Radio Garden. The former follows the sunrise across the world, streaming radio from a succession of time zones in an array of countries to bring you breakfast shows all day, every day. The Radio Garden is a goliath of a concept. Your screen fills with an image of the globe as a tide of thousands of green dots washes across the continents. Each dot represents a streaming radio station somewhere in the world and when you click on it the voices and music are there in the room, clear as crystal, wherever you are, wherever they are. There are no borders marked on the planet, no place names until you actually alight on a location: this is global radio in its purest form. When you move away from a station, until you alight on another the site makes the noise of an old fashioned radio being re-tuned. It's a brilliant way of taking radio into a digital future, while the analogue sound effects tell you the idea comes from a deep love of the medium.

While thinking about the future of radio a small discovery prompted me into thinking about the past. After my dad's death I was helping my mum settle into a new house, shifting a few boxes and sorting through stuff that hadn't seen the light of day in years. It was a strange feeling. Alzheimer's had erased him relentlessly over the last four years of his life and we watched him fade from view in front of us. Now Mum was starting a new chapter, and although he'd never lived in the new house Dad's absence was a constant. There was still a space where he should have been and reminders everywhere, in the books and photographs we removed from the boxes, and the souvenirs and mementos of a shared life now halved. It's little wonder that I was experiencing waves of nostalgia, but the postcard I found seemed to trigger something else. It was tucked between a souvenir mug from the Queen's Silver Jubilee and some of my old reports from primary school, from which I learned that I had trouble using scissors and disliked activities that involved getting my hands dirty. On the front of the card was an illustration of a pig wearing black and white checked dungarees riding a donkey that was pulling a cart in which rode what appeared to be Jack and Jill with their pail of water and Little Bo Peep. Some sort of works outing for nursery rhyme characters, it seemed. On the reverse the franking in the top right-hand corner – six and a half pence, posted in London, W1, on 15 September 1976 – was still a vivid red. Printed in brown at the lower left, using a bottom-heavy font best described as 'seventies', were the words, 'Listen with Mother'.

'Dear Charles,' the card read in neat round felt-tip, 'Thank you very much for your lovely Jumbly picture. We will hang it up on our wall.'

I can't remember drawing a lovely Jumbly picture. I couldn't even tell you what a Jumbly is, let alone whether I could accurately reproduce one to a level deserving of wall-mounting at the headquarters of the national broadcaster. I can't remember receiving the postcard and can't even remember listening to Listen with Mother, but I'm pretty sure I'd have been excited to find it waiting for me on the kitchen table when I got home from school towards the end of the stifling, drought-ridden summer of 1976. I must have been thrilled enough for my mum to keep it all these years, anyway. The fact the card was unsigned, other than the printed Listen with Mother logo, made it even better somehow. It hadn't come from a person in an office reading a name and address from the back of my scraggily scribbled crayon interpretation of a Jumbly, it had come from the programme itself. That meant it had come from the radio itself.

This postcard was the documented beginning of my radio life. My radio story started there, in the summer of 1976, hearing an invitation to send in pictures of the Jumblies on Listen with Mother, being parked at the kitchen table with crayons and paper, sticking my tongue out of the side of my mouth and sweeping great coloured lines across the blank expanse.

Holding the postcard in my hands triggered a swathe of radio memories. Being home ill from school and having Radio 1 on my bedside clock radio all day, a time when Men At Work's 'Down Under' was number one and being played practically on rotation, meaning that every time I hear it now I still experience a faint wave of nausea. Hearing my aunt and uncle on a Capital Radio programme talking about their difficulties conceiving a child – there was Uncle Phil's voice, I marvelled, coming out of the speakers, but Uncle Phil wasn't in the room. Hearing my cousin interviewed on local radio after he'd scored the winner for Millwall in the FA Youth Cup final against Manchester City. Listening to John Peel late at night, hearing 'Safety Net' by the Shop Assistants coming through my pillow from the little transistor stashed there and bunking off school the next afternoon to go to Our Price in Eltham and buy it. My dad's aircraft scanner, on which he'd listen to pilots and air traffic controllers arranging the sky above our heads. *Test Match Special* on the car radio when out with Dad, and the time both of us sat in silence in the car listening to live updates from the *Herald Of Free Enterprise* ferry disaster after he'd picked me up from football practice. *Junior Choice* on a Sunday morning. The thrill of picking up British radio in France. *The Archers* omnibus in the kitchen to the sound of carrots being chopped and meat spitting in the oven. Moving in with a girlfriend for the first time and establishing whole new domestic routines, not least that we both loved GLR, the BBC London station, its breakfast show with Jeremy Nicholas and Ruth Awbery and its brilliant weekend schedule with names like Mel and Sue, Sean Hughes, Danny Baker and Phill Jupitus. Living in Dublin and recalibrating my listening to RTÉ's *Morning Ireland*, Phantom FM, Today FM and Newstalk, finding new radio rhythms and routines (and delighting in Irish radio advertising, the cheesiest and cheapest there is, believe me).

I could almost tell the story of my life through radio, or at least *a* story of my life. We all could. Radio is such an embedded seam of our multilayered world that it's easy to forget just how miraculous and magical it is. We take this marvellous thing for granted; it's almost impossible to imagine what it must have felt like in the earliest days to have voices from London, Birmingham, Manchester and Glasgow in your house, live, as they were speaking from wherever they were. The monarch, the Prime Minister, the music hall acts, all there, in your living room, in your head. In his landmark travel book *In Search of England*, the writer H.V. Morton made a journey through the England of 1926, early in the era when radio sets were still a luxury. He reached the remote Cornish hamlet of St Anthony-in-Roseland, stayed as a guest in the cottage of a local farming couple and was invited to accompany them up the lane to listen to a neighbour's new wireless, from which the sound of a dance band playing live several hundred miles away in the Savoy Hotel, right in the heart of London, emerged. The wonderment of his hosts was palpable even if the tone of metropolitan sophistication seeping into a rural idyll portrayed as semi-feudal is a teensy bit patronising. Even so, Morton doesn't wholly approve, lamenting 'the new picture of rural England: old heads bent over the wireless set in the light of a paraffin lamp'. This was Britain in the 1920s, when a gathering in front of the wireless started to become the stock evening activity for the modern family, the peak Reithian era when announcers wore evening dress and jokes about drink, clergymen, illness and Scotsmen were banned.

That modern world according to Morton was by definition a shrinking one. For me, however, radio has always expanded my horizons. My most abiding experience of radio's magic and power came in my early teens after I was given what was grandly called a 'music centre', a combined record player, cassette deck and radio. It was a chunky, old, unwieldy thing that can't have been expensive, made cheaply by a company whose name I've neither seen nor heard anywhere else since. By this time I was old enough to stay up and listen to John Peel without the need for illicit under-pillow radio action, but before Peel's voice would ease laconically over the strumming blues of his theme music I would sit for a couple of hours cross-legged in my bedroom in front of the radio dial in a small pool of lamplight and listen to the world.

I'd attached a wire to the aerial socket and a metal coat hanger to the end of the wire and hung it on the louvre window that let in the wind, and would slowly turn the dial. Radio waves travel further at night and out of the darkness would come hiss and crackle then the echo of a distant pop song fractured into shards by the atmosphere. Pops and thumps, a faint voice too buried in static to tell which language, a burst of French, some Morse-like beeping and a sudden blast of a symphony orchestra at full fortissimo. A guttural female voice, Dutch or Flemish, I couldn't possibly know, skirled into deep, sonorous Russian as the dial turned, a faintly menacing voice of the

Cold War underlining the constant threat of nuclear annihilation. A sudden, brief rasp of static betrayed a lightning storm somewhere nearby, then there was the relief of some jangly sixties guitar pop from a pirate station, the signal rising and falling like the ship it broadcast from on the swell somewhere out on the North Sea.

I would spend most nights of the week like this, slowly turning the dial back and forth along the bands, medium and long waves, one after the other, alone in the semi-darkness but immersed in the voices, music, cracks, howls, pops and mournful moans of the world, the ether and the night. Glowing faintly on the dial in tiny letters in front of me were the names of some of the places from where these voices came. Many were familiar – Luxembourg, Budapest, Lisbon, Paris – others were less so: Lahti, Ulm, Huizen, Timişoara, Hilversum. Familiar or not, these names sounded impossibly exotic (with the possible exception of Droitwich). There I was, a bored, directionless, isolated teen in a soulless suburb in south-east London and a whole continent was swirling invisibly around me: all those voices, all that knowledge, all that music; a sampler of the world filtered through a bent coat hanger and a wire jammed into place with a paperclip. It also felt deliciously democratic progressing along a dial where Paris and Droitwich received equal billing and where Lisbon stood on a par with Hilversum.

Hilversum. The name came to almost represent radio for me: a place I'd never heard of and couldn't point to on a map, but one that seemed to symbolise the magic of those nights by the dial when voices crossed borders without impediment and one could eavesdrop on the conversations of nations. Even the name came to sound onomatopoeic: there was hiss and hum in Hilversum. It was a mythical place, part of the mixture of magic and science that made radio happen, all those words and sounds dancing and billowing in the air, sown into the skies by giant transmitter masts that were the connecting nodes of a global network of knowledge into which I, a lonely boy bound by shyness and crushingly low self-confidence, could be a part.

Finding that postcard from *Listen with Mother* opened the door to my radio life and radio world. From the gloom of bereavement I was filled with a new enthusiasm. At this fulcrum point in the history of radio I would look back and look forward. I would set out to find the magic of radio, celebrate its heritage and meet people who not only share my passion for it but who help to propagate it. The magic of radio was out there. It might even be in Hilversum, wherever that was.

Turning the postcard over in my hands as the little Roberts radio I'd just bought for Mum burbled away in the other room I decided there and then to immerse myself in the world of radio. I sensed there was magic to be found in its history and the culture of listening we all share. It would be a journey in search of stories and of people, the people responsible for making the radio world we live in today, from long dead visionaries to contemporary broadcasters. I'd move up and down the dial of history, looking for the essence of the magic that makes us all radio people, taking stock of and celebrating the medium we get for free and take for granted. I would finish the journey in Hilversum, the place whose name on the radio dial came to represent the magic of radio for me. Some people dream of going to New York, Paris or Rio de Janeiro. I dreamt of going to Hilversum, and that was just fine.

It took me a while to realise that I'd also begun to hoard old radios. It hadn't been a conscious thing but I'd started picking them up in junk shops and ordering them over the internet, not really knowing what I was looking for or looking at, but placing them around the flat, imagining the places they'd been, the stories they'd told and the people who'd gathered around them. Each radio represented a slice of intimacy, a deeply personal instrument of comfort and company just the same as my old system had been at a time when I struggled to find a place in the world, providing a life-ring in the darkest water as the sky hummed with music and voices.

2

Love Hertz: a Radio Life

The first time I appeared on the radio was around 1984 when I was 13 years old. Back then the weekday evening show on BBC Radio Kent was presented by Rod Lucas, whose programmes from that period are still some of the best, most original radio I have ever heard. It wasn't strictly a phone-in, a music show or a current affairs programme. There were bits of all three, but *The Rod Lucas Show* was simply two hours every night of whatever Lucas was in the mood for that evening. People would call in and tell a story, or a set a quiz question that other listeners could answer, tell a joke, reminisce about some aspect of local life or sing a song to him down the phone. There was little or no format: he'd play records occasionally, records of his own choosing rather than culled from some generic playlist, and what really made him stand out was the range of jingles and gimmicks he had close at hand. It was quite a few years before I twigged that when someone called in for a birthday dedication he hadn't actually got up and careered across a giant, echoing room sending stacks of crockery and piles of crates crashing to the floor to reach an old barrelhouse piano in the far corner where he'd belt out a birthday song of his own devising. Instead he was still in his chair in a small studio in Chatham, pushing and pulling cartridges in and out of a machine. I don't recall him ever having studio guests or any sidekicks next to him, there was just Lucas, his bells and whistles and the people of Kent. It was an incredible programme: anarchic, funny, original and years ahead of its time. Rod Lucas painted amazing pictures in my head and his world was one I wanted to live in.

He'd speak occasionally to a character called Weirdy Beardy who would pop up in a different part of the county each night, the location a secret until his first contribution to that evening's show. This was no slick outside broadcast from a radio car arranged weeks in advance, the Weirdy Beardy would just roll up in a town or village, find a phone box and call the studio. He could be in Tunbridge Wells on Tuesday, say, Folkestone on Wednesday, Strood on Thursday, randomly rocking up in some corner of the county and, wherever it was, whatever the weather, crowds of people would converge on the phone box to perform their party turns, ask for a dedication or set a quiz question for Lucas back in the studio.

Rod Lucas made you feel like you were his friend. He was warm, funny, and for all the anarchy and mayhem of the show, whenever he spoke to a caller he displayed a rare empathy, no matter whether that caller was eight years old or 80.

One night my friend Tim was over at our house and we decided to phone in to Rod's show. I've no idea how we settled upon performing 'Pretty Little Angel Eyes' by Showaddywaddy for the people of Kent gathering round their wirelesses for a post-prandial listen, but that's what we picked. Not just the chorus either, we're talking the whole song, harmonies and all. We ran through it a couple of times then, with Tim pacing up and down and gnashing through his fingernails as if we were about to take the stage in front of a packed Carnegie Hall, I picked up the receiver and dialled the number. I spoke to a producer, told her of our intention, she called us back and before we knew it we were on hold in a queue of other nervous listeners waiting their turn to speak to Rod. My mouth was dry, my palms clammy. Upstairs in my bedroom a blank cassette turned slowly in my tape recorder in front of the speaker. I could hear Rod's voice, joshing with Weirdy Beardy in whatever godforsaken part of Kent he'd parked himself that night, but he sounded a long way away on the other end of the line. In order not to miss our cue I pressed the receiver so hard to my ear that even today I sometimes look in the mirror and think my right ear looks flatter to my head than the left. A couple of callers went through quickly, then there was a slightly longer conversation, I think involving a question about the date of the composer Robert Schumann's wedding. Then the moment came.

'And on the line we've got Charlie. Charlie's in Lewisham,' said Lucas. 'Good evening, Charlie.'

''Allo Rod,' I blurted, my voice a-quiver but trying desperately to sound like this was no bother at all, that I performed Showaddywaddy songs to strangers down the telephone all the time. 'I've got Tim here with me and we're gonna sing you a song.'

'All right, Charlie,' he said with a chuckle, 'let's hear it, then.'

I put the receiver down on the table and retreated to where Tim was still pacing and gnawing. I counted us in, too fast of course, what with the nerves, making the first line sound more like *prililangeleyes* rather than the carefully enunciated intro we'd practised, but bless our innocent hearts the two of us stood there and sang that whole song. Verses, chorus, bridge, the lot, all *a cappella*, keeping erratic time by dipping our heads on each down beat. While we'd contrived the idea as a bit of a laugh, the further we got into the song the more serious the endeavour became. This was no longer two spotty eejits still in their school shirts egging each other on, it was a performance. We had a responsibility to the people of Kent here, not to mention to Showaddywaddy who, as far as we knew, might even be listening, to absolutely nail this thing, and the longer we sang the more serious and nuanced the performance became. Our *shoo-bops* were spot on, our *boo-dups* coming straight from the heart. With a mixture of waggling eyebrows and hand gestures we even co-ordinated the fade-out at the end of the record until our voices fell away to nothing. A moment passed. Then we remembered where we were, the spell was broken and, breathless from our musical exertions, I picked up the receiver, swallowed, and said a tentative, 'Hello?'

There was silence on the line. 'Rod?' I said, thinking Lucas might be feigning a stunned reaction to our performance, ready to shove in a cartridge marked 'concert hall ovation' while hooting and whistling his approval. But no, there was nobody there, only the faint static hiss of a dead telephone line. I looked at Tim.

'Rod's not there,' I said. We looked at each other for a couple of seconds then sprinted from the room, the receiver clattering onto the table. Having tackled the stairs three at a time we crashed into my bedroom and lunged for the tape recorder.

I scrabbled at the buttons, stopped the recording, re-wound the tape and pressed 'play'. We heard Rod say my name. We heard me announce that I had Tim here with me and we were going to sing him a song. We heard Rod entreat us to hear it. We heard us start up with the first '*prililangeleyes*', taken solo by yours truly, then the second, in which I was joined by Tim.

And then Rod cut us off.

'They're going crazy but I love it,' he laughed over our first seamless transition into falsetto. 'Well done, boys!' With that we were quickly faded out and he hailed Rachel in Rochester who wanted to know where the biggest Easter egg in the world was made, and what it weighed.

I looked at Tim, Tim looked at me. In silence I re-wound the tape.

'*Prililangeleyes, prililangeleyes.*'

'They're going crazy, but I love it! Well done, b—'

I re-wound it again. And again. We played it back countless times and with each playback our disappointment that the full weight and gravity of our performance had been denied to the county of Kent found itself being gradually eclipsed by the excitement of what we'd just done.

We'd been on the radio. Not only that we'd been on *The Rod Lucas Show*. Yes, he'd cut us off before we'd even got into our stride, but in fairness to him we'd heard as he opened the show that he wanted 'a quiet one' that night. We'd chosen to entirely disregard that notion and instead inflicted upon him two pubescent berks whose voices hadn't long broken caterwauling their way through arguably the least cool record ever to *shoo-bop* its way into the top ten. Any DJ in their right mind would have nixed that nonsense before it could spread.

Once Tim had gone home I spent the rest of the evening playing and rewinding, playing and rewinding, listening over and over again. As I did so I looked out of the window at the night sky and thought about how my voice had been flung out there under the stars. People heard it. People heard me. Granted, most of them had probably thought 'hark at this pair of twats', but no matter, I'd been on the *radio*.

The same radio through which on childhood Sunday mornings my sister and I would climb into bed with our parents where we'd all listen to *Junior Choice* on Radio 2, having tea and toast while Ed Stewart (and later Tony Blackburn) read out dedications before playing 'My Boomerang Won't Come Back', 'Puff the Magic Dragon' and 'Hello Muddah, Hello Faddah'.

The same radio I'd hear when we stayed at my nan's house, walking sleepily into her kitchen in the mornings to find the windows all steamed up and the air damp from the tea towels she was always boiling in a saucepan on the hob to get them properly white while the LBC breakfast show with Bob Holness and Douglas Cameron echoed around the walls from her tinny little transistor radio. The same radio through which Rod Lucas, Capital Radio, Radio 1 and the pirate station Laser 558 would chaperone me through an adolescence in which I struggled to establish where I fitted into the world, was bewildered by it and withdrew from much of it.

Radio helped to change that. At first it was a comfort, the soundtrack to those family Sunday mornings with tea and toast and the sudsy morning warmth of my nan's kitchen. As I got older radio showed me there were horizons beyond the railway marshalling yard behind our house, beyond the city and beyond the sea.

A little while after our radio debut Rod Lucas disappeared suddenly from the Kentish airwaves, to be replaced by someone calling himself a mock zany name, something he probably thought sounded engagingly matey – 'Jacko' or 'Jimbo', or something – but which implied an intimacy with the audience that hadn't been earned. It wasn't remotely the same.

I kept the tape of my radio debut for years but it's long gone now. It doesn't really matter as I'd played it so often I can still hear every detail. It was the moment radio stopped being a passive experience. I'd stepped briefly into a world of magic whose cartography was in the radio dial and I could travel from Hilversum to Lisbon in barely a quarter turn of a wheel.

Many years later I found myself sitting at a table covered in microphones in a studio high up in Broadcasting House, about to co-present the first in a live series on BBC Radio 4 with national treasure Fi Glover. My forays into travel writing had somehow convinced the show's producer David Prest that I would be the ideal foil for Fi on a series called *Traveller's Tree*, a combination of studio guests, pre-recorded items and listeners phoning in offering travel-related tips and suggestions. The first show was about to start. The three o'clock news was in my headphones and I watched the second hand of the studio clock tick round waiting for the moment the continuity announcer handed over to us and the microphones went live. This was a radically different situation to the Showaddywaddy incident but the feeling of nervous excitement, hearing the station output from somewhere else in the building through the headphones and the anxious, butterfly-tickled wait for the moment, was exactly the same. The previous day there'd been a production meeting in Islington where Fi and I had gone through the script with the producers and I'd spent the time in between leafing anxiously through it, highlighting my bits in fluorescent pen, committing as much of it to memory as I could.

I felt like a bit of a phoney. Actually, more than a bit, I felt a complete fraud. I'd not really served my time or paid my dues to do a show like this. I'd done a little bit of radio presenting at university and I'd spent many hours in the bowels of the BBC over the years sitting in a tiny windowless cupboard with a microphone while a stream of local stations from around the country appeared in the headphones to ask me about whatever book I'd just written, but I'd never really been an actual radio presenter before.

When I was a kid some French friends of my parents sent over a pair of walkie-talkies as a present, and I'd set one of them next to the speaker of my record player and start playing records, chuntering away between them using links I'd copied from Radio 1. Doubtless in France the walkie-talkies were tuned to an empty frequency designed for that purpose. In south-east London, however, it turned out to be one used by a popular local minicab company whose communications were drowned out by what appeared to be Dave Lee Travis on helium playing Adam and the Ants. I was oblivious to this: when I'd sign off for the evening I'd finally release the talk button and hear raised voices coming through the static featuring quite a few words I didn't understand at that age but interpreted as noisy enthusiasm from loyal listeners.

All the same, this didn't really equip me for the multi-fathomed deep end of live national radio. I woke early and anxious on the day of the show and left the house in good time – in such good time, in fact, that I could have yo-yoed between home and Broadcasting House several times and still not been late. I'd had a sleepless night of anxiety dreams featuring giant microphones preventing me from getting out of the front door, meaning that by the time I reached Broadcasting House I'd necked so much Red Bull that scampering up the outside of the building like Spiderman and slipping into the green room through a window had become a plausible option.

I emerged from Oxford Circus tube station and set out towards Langham Place wondering what on earth I was doing there. It was a feeling that wouldn't leave me for the entire three series of *Traveller's Tree* – and, indeed, all the radio I've ever done – but that first day it was particularly acute. I greeted the day's guests in the green room, hoping they wouldn't notice the wildly bloodshot eyes and persistent burping that went with the several pints of energy drink sloshing around inside me, and assured them they'd be fine, the show would fly by and they'd come out thinking they'd hardly said anything, but not to worry, they'd have said plenty. All the while I was saying this a voice in my head was whining, 'But I won't be fine, the show won't fly by and hardly saying anything sounds pretty good to me at this point.'

Then Fi breezed in and immediately set everyone at ease, even her wired, practically gibbering co-presenter, and we all made our way up the corridor, into the studio, arranged ourselves around the table and checked the microphones, asking everyone in turn what they'd had for breakfast.

Eventually a voice in the headphones said, 'A new series now on Radio 4 and here to tell you all about it – Fi Glover.' The mics went live, red lights illuminating on the table and over the door, and Fi opened with a characteristically Gloveresque, 'Hello, and welcome to *Traveller's Tree*, the show that will never ask you, "Chicken or beef?" We want your thoughts, tips and ideas, so do get in touch with the programme, and here's the lovely Charlie Connelly to tell you how' – and I was now, irrefutably, a presenter on national radio. My contribution to the half-hour show was pretty much limited to chuckling at Fi's jokes, giving the phone number and email address, reading out a few listener emails and attempting the occasional wisecrack that would fall heavily on its arse, but there I was, on the radio.

After each show I'd remove my headphones, convinced I was about to be fired for being useless, then stand up, convinced I was about to be fired for being useless, lift my coat off the back of the chair, convinced I was . . . well, you get the idea. We'd adjourn to the pub round the corner, talk about how the show had gone, and when the time came to disperse I'd realise I'd somehow survived the chop for another week. I lasted for three series, when I left the country and went to live in Ireland. Thinking back I'm still half-convinced that part of the reason for that decision was to spare the producers the awkwardness of telling me I wasn't needed any more, yet amid what had been a period of particularly low self-esteem and depression I didn't realise that radio was saving me again, if only I'd let it.

Presenting alongside Fi Glover was an extraordinary education: she marshalled the guests, the callers and the demands of a live programme on national radio during which any manner of things could have gone wrong as if she were sitting around a table at home with some old friends just talking about this and that. She made sure everyone in the studio had a fair amount of time each, made everyone feel relaxed, even me, while all the time gliding effortlessly through a script that was timed to within an inch of its life. The warmth and affection of the callers towards her was palpable too, and I thought I could never be that confident or make radio seem that effortless, the medium becoming a natural extension of my personality.

After the last programme of the third series, just before I moved to Ireland, I sat alone in the tiny green room with the discarded paper cups and the empty biscuit plate and for the first time looked up and noticed the room had a skylight. Some fluffy white cloud passed through a blue sky, but what I noticed most of all was the transmitter mast right above the room. It didn't work, it was there purely for decorative effect, but as I sat in the chair with my head tilted back watching the latticework of the mast stretch towards the heavens above me it properly dawned on me what a privilege it had been to broadcast from this building. I thought of the real, professional broadcasters who'd sat in the same studios, and how one day in a corridor I'd passed a door beside which a plaque commemorated how Charles de Gaulle had broadcast to France from there during the Second World War. I'd become part of that same story – a tiny, irrelevant part, but a part all the same.

Radio had been my comfort and solace again, only this time I'd been inside the little box of magic, and it was my voice out there in the ether, my words added to the clamour of decades, the layers of voices from the choir of British radio history that, in terms of the kind of radio we listen to today, began effectively with a remarkable contraption called the Electrophone.

3

Insinuating Eavesdropper:
the Electrophone Story

There was an astonishingly short amount of time between early technological breakthroughs in sound and the birth of broadcasting as we know it, particularly when you try to put that development into the context of human history. For a good 200,000 years we had been restricted in how far and fast we could communicate by our immediate physical limitations. From banging rocks together to waving our arms about, from lighting beacon fires to running coloured flags up poles, progress was always at a pace that made glaciation look snappy. All those millennia of advances in human development – fire, the wheel, democracy, electricity, snooker – but in communication terms we'd essentially progressed as far as cupping our hands around our mouths when we shouted.

Yet within the space of a few decades, the merest blip in the chronicle of human history, we moved from hollering out of windows at the errand boy through a rolled-up copy of the *Illustrated London News* to a human voice being heard by millions, in real time, right across oceans.

We take radio so much for granted that it's easy to forget what an impact it has made on the world. When we're drumming our fingers on the steering wheel in time to a song, or catch ourselves in the kitchen shouting at some boneheaded phone-in host, we never pause to think about the technology that's enabling us to do just that, whether we're listening via an app on our phone or a paint-spattered old transistor with a knitting needle jammed in the aerial socket. Granted, when you catch *You and Yours* or Nigel Farage's LBC show you might ask yourself why anyone bothered in the first place, but the combination of theories, experiments, processes, engineering and sheer innovation involved in bringing those voices, songs, jingles and people saying 'first-time caller so I'm a bit nervous' into your own personal, private space is utterly jaw-dropping. Especially if, like me, you're still agape at cars with electric windows.

There are radio waves all around us. Between you and the sky, you and the wall; as you read this they're swirling about between you and the page. Look out! They're in your hair. All sorts of electromagnetic stuff is oscillating around the place, and for the last century or so some of it has been piggybacked by components of words and music captured by a little box with an aerial and numbers on the top that turns those components into words and music again.

Making radio waves is easy. Take a nine-volt battery and stick a coin on it so it's touching both terminals then hold it near a radio – that crackling is caused by the radio waves you've just made (yes, it sounds pretty horrible, but probably still better than your average shock jock).

Fortunately radio waves don't all oscillate at the same frequency. There would be utter carnage otherwise: every time you went to put on a bit of Talksport your garage door would open, the microwave would explode and a passenger jet would attempt to land in your garden. The waves via which we listen to the magical medium of radio are of a lower frequency oscillation than those that make television, for example, and much lower than microwaves.

AM (amplitude modification) radio, or medium wave, as it's more commonly known, is broadcast on frequencies between 525 and 1700 kilohertz, meaning it's carried on radio waves oscillating at frequencies between 525,000 and 1,700,000 cycles per second. That sounds fast, and is indeed pretty nippy – until you compare it to the higher sound quality FM (frequency modulation) radio that broadcasts between 88 and 108 megahertz. This means that if, say,

I started a radio station and transmitted it on 99.5mHz, the non-stop eighties indie and documentaries about Charlton Athletic would be flung out into the atmosphere at 99,500,000 cycles per second, a much higher frequency, and hence much better sound quality. Also: a pretty amazing radio station.

In the relatively few decades between the earliest developments in telegraphy, whereby electrical pulses could be sent along wires in a sequence that could be decoded by a frock-coated Victorian at the other end, to meticulously pitched bass frequencies rattling the window panes via sub-woofers in crystal clear quality, where and when did broadcasting in Britain as we know it actually begin?

From the early decades of the nineteenth century a succession of crackpots, charlatans and geniuses had thrown themselves into researching and developing what became known as wireless telegraphy: the ability to send communications first via electrical impulses then using the human voice itself without the need for cables. These were remarkable people of extraordinary vision and immense creativity: British pioneers like Charles Wheatstone, David Edward Hughes and Sir Oliver Lodge; the Croatian-born Nikola Tesla; and Édouard Branly from France – to name just a few who knew their way around the mysteries of the coherer and the spark gap transmitter.

Initially, however, nobody really thought about using this revolutionary technology for fun and larks. There would be military uses, it would practically revolutionise the maritime world, and the telegraph became a tremendous boon to the railway industry, but nobody had really considered wireless telegraphy being used for any kind of high jinks. For one thing, early wireless communication was strictly person to person, not person to audience. Instead, it took the telephone to put in motion the kind of radio we listen to today.

In 1881 Paris hosted the first International Exposition of Electricity. All the good stuff was there: Edison's lightbulb, Alexander Graham Bell's telephone, and, in a copper-bottomed example of men's priorities when presented with technological advancements that could benefit the world in untold and wondrous ways, an electronic scoring system for billiards.

Also exhibiting was a 40-year-old Parisian polymath named Clément Ader who had not only already improved upon Bell's new-fangled telephone but had, in 1880, installed Paris's first telephone network. Ader would spend much of his life trying to invent the first powered aircraft, and even attracted the interest of the French war ministry. That interest was short-lived, however: when the top brass showed up for a demonstration of Ader's much-heralded third prototype aircraft they watched as the bat-winged contraption trundled along a specially constructed runway, was caught by a gust of wind, veered off into the long grass and stopped, proving nothing more to onlookers than he appeared to have invented a car as wide as a house.

Despite this setback Ader is still best remembered today for his contributions to the early development of aviation – in later life he came up with the idea that aircraft could possibly take off and land from some kind of specially adapted ship – but his contribution to the development of radio is absolute solid gold.

Clément Ader had recognised in the new telephonic technology the potential for entertainment, and at the 1881 Expo unveiled his Théâtrophone. He'd set up 80 microphones around the stage of the Paris Opera and run a cable network from there to his pavilion at the Expo, allowing visitors to listen to operatic performances live through headphones – in stereo, no less – despite being nearly two miles away from the performers themselves. Imagine the impact that must have had. It was live streaming a century before the internet. In modern terms this was practically teleportation: the listener would have thought, 'I'm here at the Expo and yet I am at the opera *at the same time*.'

Later that year Victor Hugo brought his whole family to a Parisian hotel that had installed a Théâtrophone, purely in order to try what he called the 'two earmuffs on a wall' and came away 'delighted'.

Within a decade there were coin-operated Théâtrophone booths in hotels, bars and cafés throughout Paris, offering listeners five minutes of audio magic for just 50 *centimes*. Some people even installed them at home, and by 1892 there were a hundred Théâtrophones across the city through which every evening subscribers could listen in live to a play or a concert. Marcel Proust had one at home, by his bed, so he could swoon through his frequent illnesses to the sound of a new Debussy opera.

It wasn't long until someone had the bright idea of doing something similar in London. The Electrophone must have seemed like witchcraft when it was introduced to the UK in 1895 by Mr H. Booth, head of the newly incorporated Electrophone Company Ltd. Capitalising on the success of the Théâtrophone and the increasing proliferation of the telephone across London the Electrophone was effectively a souped-up version of Ader's system – a subscription service with its own dedicated operators at switchboards through which telephone owners could listen live to plays, concerts and church services without leaving their homes.

Using a set of earphones in a rigid contraption shaped like a pitchfork, held under the chin and connected to their telephone line, the subscriber would ask the operator to be put through to Electrophone HQ where, standing at a vast manual switchboard in London's Gerrard Street, the company's operators were ready to connect callers to the event of their choice. Microphones were in place among the footlights of stages in the city's swankiest theatres and concert halls, the odd music hall and in some churches, where they were even disguised as bibles in order not to subvert the ecclesiastical atmosphere with technological intrusion.

By 1908 Electrophone connected 600 subscribers. There were 1,000 by 1919, and the service peaked at 2,000 in 1923. Electrophones were provided free of charge to hospital patients, and during the First World War they were installed at military hospitals for use by wounded soldiers. If you could afford to have one at home the Electrophone offered pretty decent value for your £5 a year (roughly £150 today): there was no installation charge, you had unlimited usage of the system and even your calls to Electrophone didn't cost you anything.

The Electrophone was a remarkable thing in its day. We take for granted our 600 television channels and the world's radio a click of a button away, but for people for whom home entertainment had meant a couple of hands of whist and occasionally poking the fire, suddenly they had to decide between a Shakespeare play at the Aldwych or a Beethoven symphony at the Albert Hall, or perusing the list of Sunday church services trying to pick one sure to have a whizzbang sermon packed with zinging one-liners. Electrophone soirées became fashionable among the London elite, with multiple headsets available for friends and a ready supply of cigars and sherry. Queen Victoria herself was reportedly a fan after enjoying a performance of 'God Save the Queen' sung to her at Windsor Castle over an Electrophone from the stage of Her Majesty's Theatre on Haymarket on the occasion of her 80th birthday in 1899. Indeed, such a status symbol did it become that accommodation rental advertisements would cite the Electrophone as an added perk. In 1922, for example, a four-bedroom, two-reception apartment in Hampstead with telephone and Electrophone could be had for £41 per week.

'Large numbers of persons imagine the Electrophone to be a phonograph in which one hears what is apparently the voice of a corncrake with a bad cold,' wrote one enthusiast in 1900. 'Some take it to be a telephone; others are under the impression that it is a musical instrument. It is none of these things. It is simply a wonderful apparatus of Anglo-French origin which transmits sounds exactly as they are produced. Not to mince matters, it is an insinuating eavesdropper.'

The peak Electrophone year of 1923 was also, however, a big year for radio. The year-old BBC was getting into its stride and the Post Office was being swamped with applications for what were then called 'receiving licences'. Even accounting for the growth of radio, however, the demise of the Electrophone was surprisingly swift. Within a year of that subscription peak the number of Electrophone subscribers had fallen by half and by the summer of 1925 the Electrophone Company had seen the writing on the ether and packed it in, going into voluntary liquidation. A relatively brief, slightly odd golden age of cable telephony was over.

Well, not quite. Like the proverbial Japanese soldier still fighting the Second World War years after it had finished, two subscribers in Bournemouth, a Mrs Cooper and a Mrs Hatchcock, clung on to their Electrophone service until 1937. Twelve years after the company had closed down and eight years after both of Bournemouth's participating theatres had disconnected their lines, two women refused to believe it was over as long as they could tune in to their local church.

Bournemouth was the only town outside London to boast an Electrophone service, mainly because John Holt lived there. Not only had Holt worked as a telephone engineer since the earliest days of the medium, he had set up the first experimental Electrophone link for H. Booth in London in 1883. The occasion was the 41st birthday party of composer Arthur Sullivan, at the climax of which the birthday boy, with great ceremony, lifted the telephone receiver Holt had installed and was immediately connected to the Savoy Theatre, where a production of his and W.S. Gilbert's comic opera *Iolanthe* had just concluded. The cast gathered round the microphone, set up by Holt, and ran through a selection of the opera's highlights as the receiver was passed around the room among Sullivan's guests, including the Prince of Wales who, as King Edward VII, would make great use of the Electrophone at his royal residences.

Before that, in 1880, Holt had installed the first telephone in a British church, in Childwall, Merseyside, for the benefit of a parishioner who lived a mile away but couldn't travel to services any more. All of which might explain why in 1937 four Bournemouth churches were the only premises in the country still maintaining their Electrophone equipment: thanks to Holt dutifully keeping the system running they could compete for the remote worship of two ladies who had no truck with that new-fangled wireless. With the obliging Holt circulating between the churches and the homes of the last two Electrophone users in the country, fixing a loose wire here, servicing a microphone there, there was no need for Mrs Cooper or Mrs Hatchcock to invest in one of those confounded new radio contraptions.

In 1938, however, the Post Office noted that there were no longer any Electrophone users to be found anywhere in the country. Bournemouth's Electrophone had fallen silent for the last time because its long-serving engineer had also fallen silent. John Holt, who had single-handedly maintained the last vestiges of an ingenious system once patronised by royalty, died in December 1937 and the age of the Electrophone died with him. He was 93 years old.

4

Follow Spot to Light Icing: an Afternoon with Corrie Corfield, Part One

When Corrie Corfield invited me to sit in with her on a Radio 4 announcing shift there was an odd sense that I was meeting someone for the first time yet also greeting an old friend. She appeared at my shoulder in the reception at New Broadcasting House while my bag was being put through the security scanner, and it was slightly strange at first to hear such a familiar voice coming out of an actual person rather than through the grille of a radio speaker.

If anyone has a claim to being the voice of a network it's the continuity announcer. At no other point does it feel more as though the station itself is talking to you than when the links are given between the actual programmes. Corrie Corfield and her colleagues in continuity are the true sound of the BBC.

With that lofty status comes a heavy weight of responsibility. Stuart Hibberd, who joined the BBC as an announcer in 1924 and became the first representative voice of the corporation, remaining so until his retirement in 1951, said, 'One has to be to a certain extent impersonal but never inhuman.' Hibberd had relayed some of the nation's most significant moments of the early twentieth century: the General Strike of 1926; the death of King George V a decade later, before which he'd memorably prepared the nation for the worst with the update, 'the King's life is drawing peacefully to its close'; and the newsflash that notified the country of the death of Hitler. Hibberd was a master of the controlled silence. 'This . . . is London,' he would begin, while at the end of the broadcasting day he would close the station down with, 'Goodnight everybody . . . goodnight,' the final pause designed to allow the listener to respond with their own 'Goodnight.'

With that late-night pause, in particular, Hibberd, arguably the most popular BBC announcer of all time, established the inherent intimacy of the announcer's role, yet he also managed to combine human warmth with the retention of an unimpeachable air of authority. He was probably most responsible for the perpetuation of 'received pronunciation' becoming the BBC standard, an issue he acknowledged himself around the time of his retirement when he admitted that broadcasters like him were 'very conscious we are using a lot of southern English, generally speaking, and owing to the diversity of people who inhabit this island our speech must be anathema to hundreds of thousands of our fellow citizens'.

These days the BBC's newsreaders and continuity announcers have a much wider range of voices, from Kathy Clugston's lilt from the north of Ireland to the sonorous mellifluousness of Neil Nunes and his Caribbean-rooted tones.

Corrie escorted me through the revolving doors of New Broadcasting House and gave me a whistle-stop tour of the gargantuan news operation on the ground floor. Then it was up in the lift and along a corridor into the original part of Broadcasting House and the slightly smaller, quieter operation that comprises Radio 4's continuity department – a small office occupied that day only by the continuity producer, just along from the studios. We had a bit of time before the midday shipping forecast so I asked Corrie how on earth someone came to be a Radio 4 continuity announcer. I don't remember seeing any leaflets about it in my school careers office otherwise, who knows, I could have been the first Radio 4 announcer to speak with a cockney-tinged nasal whinny, and that's a loss the BBC will just have to rue now.

'I joined the BBC originally as a trainee studio manager, which is a bit like a sound engineer,' Corrie told me as she printed out a copy of the shipping forecast for me. 'The BBC used to have lots of training courses back then in the early eighties and there were three studio manager courses each year, with an intake of 12. In the summer of 1983 I was working at the Mermaid Theatre in Blackfriars as a followspot operator on a production called *Trafford Tanzi*, a play about a woman wrestler that starred Toyah Willcox. I'd studied drama and English at Goldsmith's University but came out not knowing what I really wanted to do. I was quite keen on the BBC, and I had already applied to them when I was still at school to be a trainee secretary.'

Corrie's headmistress had sat her down and told her she'd most likely regret passing up the opportunity to go to university. She applied to the BBC anyway and was given an interview, where she was told she was exactly what they were looking for but that she would probably regret not going to university. She went to university.

'So there I am, fresh out of Goldsmith's and working at the theatre, wondering if my destiny is to become a stage manager, when one night the sound engineer I was working with pulls out an application form and starts filling it in. I was sitting next to him having a cup of coffee and asked what it was and he explained it was for a studio manager's course at the BBC. "Oh," I said, "that sounds interesting, let's have a look." It turned out he had a spare form, so he handed it to me. I filled it in and sent it off and, well, I ended up getting in and he didn't.'

The studio manager's course was the launching pad for a range of different careers within the BBC and proved a fertile breeding ground for talent of all kinds.

'Years later I meet so many people at the BBC who started as a studio manager but went on to do different jobs,' said Corrie. 'If you look at some of the announcers here of a similar age group to mine – Neil Sleat, Jane Steel, Diana Speed – we're all ex-studio managers who became announcers because that scheme was such a good way to become steeped in every aspect of radio. By the time I came to be an announcer I'd listened to so much and made so much radio I was absolutely infused with it. The course doesn't exist any more, alas, and for the last ten years it's been noticeable when we're looking to recruit announcers that there's no longer that production line of people coming through.'

Corrie began working in production for the World Service at Bush House – a 'knob twiddler' as she called it – working on the Bengali and Latin American services. She was yet to sit at a microphone, indeed the possibility hadn't really crossed her mind. 'Then I had a six-month attachment to the World Service as an announcer and a newsreader and it was while I was there that an attachment came up here at Radio 4,' she said. 'The BBC did that a lot in those days: you'd go somewhere for six months, and with a bit of luck and if you'd liked it there'd be a job at the end of it. If for whatever reason it hadn't worked out you still had a job to go back to wherever it was you were before. In 1987 we launched this sort of offshoot of the World Service called BBC 648 on the medium waveband, where you could hear the World Service in the UK and Europe from the transmitter at Orford Ness in Suffolk. That frequency had originally carried the German and French language services which would just cut in and out of the regular World Service programming that was going out on its other wavebands, which meant you could be listening to the World Service and suddenly without warning you'd be hearing programming in German or French. There was no announcement, no continuity, just someone throwing a switch somewhere and taking over the transmitter.'

BBC 648 launched on 9 May 1987, Europe Day, designed to remove that clunking gear change between different transmissions, and also to have a BBC radio service more focused on countries in the European Union.

'648 was pretty hairy,' Corrie recalled. 'At the World Service people would say, "You do realise 300 million people are listening to this news bulletin, don't you?" Which was very difficult to get your head around. The stories that would come in from around the world too: in the first news bulletin I ever had to read, a straight read of nine minutes, there was a report of a coup that had taken place in Fiji, and it had involved someone called Ratu Sir Penaia Ganilau, a name I've never forgotten to this day because I spent ages walking up and down outside the studio chanting "Ratu Sir Penaia Ganilau". Of course, by the time the bulletin was actually ready to go the coup was over, everyone had gone home and I never got the chance to say "Ratu Sir Penaia Ganilau" on air.

'I hadn't been at 648 long when there was the big storm one night in 1987 and I was on duty with a woman called Pamela Creighton, who must have been there during the war – an extraordinary old bird, one of those fabulous women broadcasters who had been there from the year dot and never once taken any prisoners. She smoked Capstan Full Strengths, and I remember halfway through the news bulletin, as the storm started raging outside, suddenly all the lights went out. She took out her lighter, sparked it up and read the rest of the script by the light of its flame. I could see in her eyes that she was absolutely loving it, it was like being back in the war. All she needed was her ARP helmet and she'd have been properly in her element.'

The World Service also threw up challenges unique to the network, challenges that probably made 'normal' BBC announcing seem pretty straightforward afterwards.

'You had to be very precise in continuity there because you had what were called switching pauses: you'd speak for 11 seconds then shut up for four seconds while some relay station in Malaya or somewhere threw a switch that pushed you on to another place in the world. You also had to be really on your toes with language too. You could never refer to "today" or "tomorrow" because you were broadcasting to so many time zones. I also did some announcing on the schools network when I first joined Radio 4: *Music and Movement*, that kind of thing, a couple of hours here and there that proved invaluable experience in front of the microphone.'

With the BBC under ever-greater scrutiny over how it spends the licence fee, and with funding being constantly trimmed, these services seem like indulgences today, even if you could argue they are simply fulfilling the BBC's role as a public service broadcaster. More and more of the BBC's less mainstream services will disappear as long as the corporation feels it has to be seen to be competing in a marketplace rather than utilising its public funding to best serve its public. It also means fewer opportunities for talent to emerge and flower away from the mainstream spotlight.

'There was time to develop people then,' said Corrie. 'On Saturday afternoons, for example, Radio 4 used to stay on FM but on long wave you'd have things like the Open University, the perfect place for a new announcer to gain experience and confidence. I think it's harder for people now because you're thrown in at the deep end a bit more, which isn't ideal for either the announcer or the listener.'

Corrie picked up the midday shipping forecast she'd printed out and annotated with timing cues and led me along the corridor into a small, dimly lit studio. She sat at the console, put on a set of headphones, laid the forecast in front of her, pushed some buttons, exchanged messages with a producer to make sure everything was working and in order, and waited to commence one of the oldest and most celebrated items in the broadcast schedules.

5

'Romantic, Authoritative, Mesmeric':
the World of the Shipping Forecast

'And now the shipping forecast issued by the Met Office on behalf of the Maritime and Coastguard Agency at 11.30 on Saturday 4 November. There are warnings of gales in Lundy, Fastnet, Irish Sea, Shannon, Rockall, Malin, Fair Isle and Faeroes.'

The shipping forecast is one of our finest national achievements, and I am prepared to fight anyone that disagrees, shirts off, in a barn, with a paying audience, for a share of the take and some meat. Watching Corrie commence the two-minute reading, the shortest of the four briny meteorological recitations that go out each day, I thought about how and why the shipping forecast showcases so much that's worth celebrating. Before long I was in a full-on reverie. For a start, when you're as much a fan of the shipping forecast as I am, actually being there in the studio as it goes out is a little like being in Sun Studios in 1954 when Elvis was knocking out 'That's All Right Mama', or seeing the Sex Pistols at the Manchester Free Trade Hall. Only quieter.

The shipping forecast is the product of drive, vision and the cocking of snooks at authority for the greater good. It's a pure form of altruism. There's no profit to be made from the shipping forecast; its credit column is calculated in saved lives, and even that figure can only be guessed at.

It's the pinnacle of writing in the English language: concise, lyrical, rhythmic, poetic and evocative. It has strict rules of composition and vocabulary – no more than 350 words, 370 for the late-night forecast that includes far-off Trafalgar, an area deemed close enough for inclusion in one daily forecast but too far away for all four – yet still manages to release the imagination from the confines of its litany. Essentially the shipping forecast is a national epic, like *Beowulf* or *The Canterbury Tales*, only it begins at our borders, looks outwards rather than inwards and changes four times a day while always remaining the same. The cast list is the same 31 characters each time, starting with Viking, ending at South-East Iceland, and they're represented by combinations of the same numbers and phrases.

The shipping forecast reflects and reminds us of the fact that we are and will always be a maritime nation and binds us to our continent, its map covering an area that encompasses the coast of Europe from Norway all the way round to Portugal, surrounds Ireland, and even reaches all the way up to Iceland. It's a reminder that the seas around these islands are not a moat but a link, a hub, a shared space connecting us with nations as large as Germany and as tiny as the Faroe Islands. It's geographically democratic: Thames carries the same weight as Fair Isle, the little Severn Estuary island of Lundy as the Bay of Biscay. It's classless too, aimed as much at the two old lads in their fishing boat hoping to catch enough herring to sell at the quayside to get them through another day as the cruise liner captain or the stockbroker on his yacht. Everyone is equal in the eyes of the sea and everyone is equal in the eyes of the shipping forecast. In the truest spirit of altruism, all lives here are worth the same. The shipping forecast helps us to never forget this.

'The general synopsis at 06.00. Low Faeroes 987, expected Norwegian Basin 991 by 06.00.'

The shipping forecast somehow turns names and numbers into poetry. There's an innate rhythm to it: it scans, it lilts and evokes. It's inspired plenty of poetry, from Seamus Heaney in his *Glanmore Sonnets* to

Carol Ann Duffy's 'Prayer' via Seán Street's wonderful 'Shipping Forecast, Donegal'. Radiohead and Blur have featured it in songs, Tears For Fears' 'Pharaohs' is one of the finest uses of the forecast in popular culture, while the DJ Overseer beautifully intertwines a forecast read by Brian Perkins with a deteriorating state of mind in his ambient track 'Heligoland'. Alec Guinness, who when you think about it had the perfect voice for reading the shipping forecast, called it 'the best thing on the radio; romantic, authoritative, mesmeric'.

We never know the names of the shipping forecast authors at the Met Office. They sit there in front of their banks of computer screens with their ticking wind arrows and rings of isobars bulging and contracting in rainbow colours, faithfully double-checking the wind speeds expected in North Utsire and achieving levels of the meticulous that led to me hearing one morning the gloriously compass-complete 'north-west South-East Iceland'. They sit there in an office in Exeter swinging in an office chair, apple core browning in the wastepaper basket, somehow contriving to produce some of the most beautiful pieces of writing in our language not just once but four times a day, every day, writing that blurs the lines between functional, prose and poetry, inspiring our great poets to some of their best work and, most importantly, causing the salty old seadog listening in his wheelhouse to decide against going out, after all, for there is a gale coming.

'The area forecasts for the next 24 hours. Viking, west veering north-west three or four, increasing five or six, showers, moderate or good.'

Like a clue from a cryptic crossword, what on the face of it seems nonsensically impenetrable is actually fairly straightforward to understand once you know what you're looking for. The format is the same for every area, every forecast: first there's the wind, the direction from which it's coming and its predicted strength on the Beaufort Scale, followed by the weather conditions and, finally, the levels of visibility.

The forecast progresses in the same order of areas, starting in the north-east with Viking and progressing broadly clockwise, zigzagging across the North Sea, scooting down the Channel, turning south towards the Iberian peninsula and racing up to South-East Iceland. It's broadcast four times a day, at 00.48 and 05.20 on Radio 4 long wave and FM, and at 12.01 and 17.54 on long wave only, except at weekends when the 17.54 forecast also goes out on FM.

'North Utsire, South Utsire, south-westerly veering north-westerly five or six, occasionally four, rain then showers, good, occasionally poor.'

I have been to Utsira, and there aren't many people who can say that. The shipping forecast inexplicably spells the two sea areas differently from the name of the island itself – 'Utsire' was the official Norwegian spelling until it was changed to Utsira in 1924, yet the island didn't appear in the shipping forecast until 1984 – but the spelling discrepancy didn't prevent me from finding the place at the end of a heaving 70-minute ferry ride from the Norwegian port town of Haugesund across some of the most notoriously tempestuous seas in northern Europe.

Just over 200 people live on Utsira – the islanders are called *Sirabu* – in settlements spread along a valley between two rocky promontories overlooked by a lighthouse. It's rugged but beautiful, populated by a friendly community who were slightly baffled by the arrival of someone who had travelled there because their island is mentioned in a British weather forecast. Generally the only tourist visitors are birdwatchers; the island has the largest variety of birds to be found anywhere in Norway. Other than its eight mentions a day on BBC radio Utsira has only really popped its head above the global parapet once, when in 1925 the island's midwife Aasa Helgesen became the first woman mayor in Norwegian history.

I stayed in a cabin in a remote spot on the northern shore, and on my first night tuned in to the 00.48 broadcast, which is quite possibly the most shipping forecast way of listening to the shipping forecast you could possibly devise.

'Forties, south-west, veering north-west four or five, increasing six or seven later, rain then showers, good occasionally poor.'

'When I'm ancient, and boring people to death about my younger days, I'll still be talking about the shipping forecast,' Corrie had told me as we'd sat waiting for noon to come around. 'Late at night when you're at the end of a long shift and still facing the extended 00.48 forecast sometimes you think, "Oh God, do I really have to speak for ten minutes straight now?" But that really doesn't last long because it's so beautifully written and such a joy to read. Every single word is important because every single word has been pored over to ensure there's nothing superfluous or spare.

'I read a special recording made for Judi Dench's *Desert Island Discs* because she wanted a shipping forecast that included Finisterre as one of her choices. As it happened I'd read the very last forecast before Finisterre became FitzRoy in 2002 – because the Spanish shipping forecast also has an area Finisterre that's slightly different to ours – so they asked me to re-record it, which I thought a great honour.'

'Cromarty, Forth, Tyne, north-west four or five, increasing six or seven, perhaps gale eight later, showers, good.'

'Sailing By', the theme that precedes the 00.48 shipping forecast, was written by the British light music composer Ronald Binge. Originally composed in 1963 it first accompanied film footage of hot air balloons, but since 1978 it's heralded the late-night broadcast of the shipping forecast. It's been selected as a Desert Island Disc by Jarvis Cocker and Michael Ball, and most recently featured strongly in the Ken Loach film *I, Daniel Blake*.

In 1993 the BBC stopped playing *Sailing By* before the shipping forecast as a cost-cutting measure (royalties were paid to Binge's widow for every usage). The corporation resisted public pressure for more than a year before conceding and returning the piece to the airwaves in 1995.

'Dogger, cyclonic, becoming north-west four or five, increasing six at times, rain then showers, good occasionally poor.'

The weather terminology of the shipping forecast never resorts to hyperbole. The shipping forecast doesn't have opinions. It doesn't even have adjectives. Rain is just that: 'rain'. It's never 'heavy rain', 'lashing with rain' or 'practically a monsoon out there, lads'. The weather in the shipping forecast is never 'treacherous', 'atrocious' or 'dreadful'. There might be 'precipitation within sight' or some 'light icing', but rarely anything more adjectivally daring than that, all of which gives the thankfully rare declarations of 'violent storm 11' and 'hurricane force 12' more impact when they arrive.

In 2009 BBC radio investigated the phenomenon by which many of us struggle to remember the general weather forecast almost immediately after we've heard it broadcast. Researchers in the Netherlands found that their radio weather forecasts accrued barely 30 per cent recall among listeners afterwards. The Radio 4 programme *PM* conducted a few experiments, asking the weather presenters to read their forecasts in regional accents, with a bed of soothing music beneath them, and even with sound effects appropriate to the kind of weather they were predicting. None of it worked until Peter Gibbs read out the regular weather forecast in the style of the shipping forecast. That went down very well indeed.

'Fisher, German Bight, south five or six, becoming cyclonic, then west four or five, rain then showers, good, occasionally poor.'

When I spoke to Charlotte Green, one of the great shipping forecast voices, I was surprised to learn she hasn't read one for 20 years.

'Back in 1998 Brian Perkins, Peter Donaldson and I were chosen as the three permanent newsreaders, so I stopped reading it then,' she told me. 'It's still something I'm associated with though, and I'm quite happy for that to be the case. Simon Hoggart, whom I knew through the *News Quiz*, once wrote an article in the *Guardian* headlined 'Charlotte, the Fisherman's Friend' and that stuck. I even used to get Valentine cards from sailors. They'd write me poems that I still have in shoeboxes in my study. Some of them were quite creative.

'The shipping forecast is the closest I ever came to reading out poetry on the air. There's a real appeal about it, even if you don't know anything about the sea and don't know what the terms mean; people still find it absolutely mesmerising. I find it fascinating that people with no connection to the sea are so attached to it.'

'Humber, Thames, south, veering north-west four or five, occasionally six at first, rain then showers, moderate or good.'

New Year's Day, 1924. Wreckage from the French airship *Dixmude*, which had exploded over the Mediterranean just before Christmas with the loss of all 52 people on board, began to wash up on the coast of Sicily. Rain-deluged Paris was on flood alert as the level of the Seine inched further upwards with each tide. Laszlo, a noted Hungarian spiritualist, was unmasked as an old fraud (the 'spirit faces' that manifested themselves during his séances turned out to be goose fat smeared on cotton wool). A ban on pipe-smoking came into force in Bulgaria.

At nine o'clock on that cold, wet, foggy London morning, from a room inside Adastral House, the headquarters of the Air Ministry at 1 Kingsway in central London, the very first shipping forecast was broadcast.

It was called *Weather Shipping*, was produced by the Met Office 'in appreciation of the valuable help given to the meteorological service of this country by the weather reports from ships' and sent not via the BBC but through the Air Ministry's own transmitter, flinging it far out into the storm-spun North Atlantic at a range of up to 2,400 miles. There was another *Weather Shipping* broadcast at 8 p.m., and thus began a daily ritual destined to become a much-loved and fiercely protected iron horse of British popular culture.

Gale warnings had first been transmitted in Morse code to the north-eastern Atlantic in 1911 but, after their suspension at the outbreak of the First World War, it wasn't until 1921 that maritime weather forecasts were sent out again, from the Marconi wireless station at Poldhu, Cornwall, with barometric pressure readings and wind speeds issued in code. From October 1925 *Weather Shipping* was broadcast via the new powerful BBC 5XX transmitter at Daventry, read once at normal speed and then repeated at dictation speed to enable ships' captains and radio officers to note down the necessary details. The forecast fell fully under the auspices of the BBC's output early in 1926 and was first broadcast as part of the BBC's schedule on Sunday 24 January at 9.10 p.m. The term 'shipping forecast' appeared for the first time in the *Radio Times* a week later.

'Dover, Wight, west five to seven, decreasing four for a time, showers, good.'

For all the affection, romance and cosiness that surround the shipping forecast, it exists solely because of the inherent danger of the seas. For all the songs and poems it has inspired over the years, one of the most effective and appropriate uses of the forecast in popular culture came in a 2013 drama series on Channel 4 called *Southcliffe*, concerning a fictional mass shooting in a town on the Kent marshes. There was a captivating atmosphere to the series, a malevolent stillness in which the sparse, clipped dialogue was allowed to hang. Filmed in and around Faversham, *Southcliffe* captured the foggy flatlands and marshes of the estuary in a way that was brilliantly evocative and almost spitefully unsentimental. This was no postcard depiction of the British coastline; the fog and the silence created the perfect backdrop to the unfolding individual stories of the lonely and the lost. It must have come as a surprise to some when the shipping forecast featured strongly in each episode.

We place a benign nostalgia around the shipping forecast, especially those of us who don't work on or by the sea. The lilting, rising and falling scales of Ronald Binge's 'Sailing By' have a warmth about them that compounds the much-loved poetic litany of sea areas, coastal stations and inshore waters. So what was the shipping forecast doing in a full-on, blood-spattered, edge-of-the-seat, emotionally eviscerating television drama? This was a daring subversion of the place we perceive the forecast to have in our society. It was incredibly effective because for all our sentimentalising of it the shipping forecast is essentially about tragedy, or at least attempts at its prevention.

Wherever you go on the coasts of these islands you'll find memorials to disaster at sea; plaques and stones, statues and bas-reliefs, commemorating the loss of trawlermen or a lifeboat crew, the passengers of a sunken liner or a lone fisherman. The shipping forecast, for all our blowing-the-steam-off-a-mug-of-cocoa nostalgia, is in reality inextricably linked to tragedy and death. I found it notable that when the forecast was used in *Southcliffe* it was always as the background to a person at their lowest ebb, whether it be the soundtrack to a man heading into the early morning mist to kill at random or a journalist waking up alone to the realisation he's lost both his job and his family.

Such deep-set introversion is wholly in keeping with the shipping forecast. The broadcast is stripped back and raw, you hear just one voice whose tone and rhythm is unchanging, the roll call and pattern of the places and information immovable and resistant to change or whimsy and, despite the sentimentality with which it's been doused over the years, the shipping forecast is pointedly unsentimental. We forget sometimes that the shipping forecast is there to save lives. For all the tea towels, framed prints and cosy nostalgia with which it's been festooned over the years, the actual purpose of the forecast is as deadly serious as it is vital.

'Portland, Plymouth, north-west five to seven, perhaps gale eight later, showers, good.'

Initially there were only 13 sea areas: Shetland, Forties, Tay, Dogger, Humber, Thames, Wight, Channel, Severn, Mersey, Shannon, Clyde and Hebrides. The forecasts were suspended for the duration of the Second World War, during which meteorological technology advanced to the stage where upon resumption the forecasts could be far more detailed. The sea area map was redrawn in 1949 to look almost exactly as it does today, the 13 areas divided up and doubled to 26, extending right up to Iceland for the first time.

Since then there has been only minor tinkering. In 1955 Heligoland became German Bight and Iceland became the more specific South-East Iceland, while Fisher, Viking and Trafalgar were introduced. North and South Utsire were added to the map in 1984 as part of a general revision of maritime forecasts among the nations bordering the North Sea, while the last significant change came in 2002 when area Finisterre became FitzRoy.

'Biscay, north-west three or four, increasing five or six, thundery showers, good.'

The most popular of the forecasts seems to be the late night one at 00.48, around which gather the nation's insomniacs. For all the lateness of the hour it's the one Radio 4's announcers seem to prefer as they can relax into the 12 minutes available and ease their way through the area forecast, the inshore waters forecast and the weather reports from coastal stations. It's probably the cosiest and most intimate of them all, with most of the listenership tucked up in bed and thankful they're not out in a dark and forbidding Bay of Biscay having wintry showers flung at them by a force-eight gale.

It's the last thing broadcast on Radio 4 before it hands over to the BBC World Service for the night so there's a real sense of conclusion, of the day being officially over once the last coastal station report is sent out into the night.

My favourite, however, is the early morning forecast that goes out at 05.20. Indeed, in our flat the radio alarm clock wakes us just in time to hear Radio 4 pick up the reins again from the World Service and commence the day with the shipping forecast. There's a reassurance about being slowly roused from sleep by the familiar rhythmic intonation. There's no panic, it's a gradual process, the sea areas and their solemn weather forecasts curling around the last wispy fragments of a dream, and then the weather reports from coastal stations, sentinels who've kept a night watch on the seas and are informing you just how far they've been looking: 'Stornoway, west by south, two, recent snow, 19 miles, 1018, now falling.'

'Channel Light Vessel Automatic, north by west, five, 11 miles, 1019, falling slowly.'

By the time the inshore waters forecast comes around I'm awake enough to set off on its jaunt around the coastline: 'Cape Wrath to Rattray Head including Orkney . . . North Foreland to Selsey Bill . . . Mull of Kintyre to Ardnamurchan Point.' It's as if I'm being taken by the hand, shown around the nation's extremities and reassured that everything's where it should be and as it should be. When we've made it to Shetland the day can start in earnest and all is well.

'Fitzroy, north-west five to seven, occasionally four later, rain then showers, good, occasionally poor.'

One day in August 2017 I received a call from a radio station asking if I'd go on air to talk about the 150th anniversary of the shipping forecast. '*The what?*' I asked, to be told by a researcher that the Met Office was claiming the following day as a significant shipping forecast landmark: 150 years since weather forecasts had been published in the press daily. As anniversaries go it was a pretty flimsy premise at the best of times, but its purported link to the shipping forecast, the more I thought about it, lifted me from slight bewilderment to a dudgeon of increasing altitude.

Why did I become irrationally exercised about what was essentially a harmless bit of PR cooked up in a meeting? It was because of Robert FitzRoy, the father of the shipping forecast, the weather forecast and one of the greatest figures this country has ever produced – a man who's been criminally underappreciated for too long for the countless thousands of lives that have been saved at sea thanks to his extraordinary drive and energy in the face of personal frustration, not to mention the antipathy to bureaucracy that would drive him to his grave.

To their credit the Met Office had tried to hail FitzRoy as the hero he was, from restoring his neglected grave in south London to renaming sea area Finisterre in his honour in 2002. Back in the mid-nineteenth century, however, FitzRoy was grossly mistreated by the Met Office's predecessors at the Board of Trade, something I felt was being compounded by this spurious anniversary.

In 1854 Robert FitzRoy, already notable for being the captain of the *Beagle*, with whom Charles Darwin sailed on the voyage that would produce *On the Origin of Species*, was put in charge of a fledgling Meteorological Office as a result of a pan-European convention on weather held in Brussels. As international sea trade and transport boomed, more and more ships, people and cargoes were being lost for want of reliable weather information, and there was a Europe-wide desire to do something about it. Britain, being Britain, considered itself above all those European johnnies and so only sent delegates at the last minute – so late that they missed the first day altogether – with instructions not to agree to anything. The benefits of this pan-European co-operation were so obvious, however, that the delegates couldn't find any reason not to sign up. So humpty was the Board of Trade at having to go along with this straitening Eurocracy that when FitzRoy was appointed to put Britain's meteorological department together they wouldn't even give him an office. He ended up having to rent a room at his gentlemen's club.

FitzRoy worked tirelessly from the start to achieve near miracles in the collection of weather data and, by 1861, felt confident enough to predict when storms that might put lives in danger at sea were approaching. Advances in telegraphy meant weather reports from the 15 coastal stations he'd opened around Britain could reach him quickly too, meaning that, on 6 February 1861, FitzRoy was able to introduce an ingenious system of cones suspended from frames at points around the coast, whose presence warned ships of impending storms and the direction from which they were approaching. Now, if any date has a claim to being the first shipping forecast that one is pretty watertight. It also serves to illustrate how the entire history of the weather forecast we know today essentially revolves around the shipping forecast: all the research, all the data-gathering, even the simple barometers FitzRoy devised to suit the most basic vessels, all of it was about saving lives at sea.

Within six months he was confident enough in the data he and his small team were compiling to actually predict the weather for the ensuing 24 hours. On 1 August 1861 he published, alongside the data tables from his coastal stations that he already inserted into *The Times*, the first weather forecast: a general, three-line forecast for the north, the west and the south (another strong candidate for the very first shipping forecast).

The *Times* forecasts ran for the next four years or so. Given the limited technology available to FitzRoy their accuracy was a bit hit and miss, prompting a range of stuffed shirts, from the Board of Trade to editorials in *The Times* itself, to snigger into their sleeves at a man who was undeniably achieving extraordinary things.

Eventually FitzRoy, troubled by depression for most of his life and frustrated that people were still dying in storms at sea despite his herculean efforts (when a ship called the *Royal Charter* went down in a storm off Anglesey in 1859, costing hundreds of lives, he took it particularly badly), locked himself in his dressing room one Sunday morning in 1865 and cut his own throat.

Almost immediately the Board of Trade sensed the opportunity to save a few quid and had FitzRoy's hardworking staff stopped from issuing any more forecasts. His suicide also ensured his legacy would, in the prevailing mentality of the day, be tarnished – even his funeral was a heartbreakingly miserable, sparsely attended affair when it should have been a national tribute.

Unsurprisingly there was a subsequent increase in maritime losses – of ships and lives – until the weight of public opinion forced the powers that be to reintroduce the issuing of weather forecasts in August 1867. And that's what the Met Office were claiming as the shipping forecast's 150th birthday: the grumpy, reluctant capitulation of their meteorological predecessors who had obstructed and belittled the tireless work and achievements of the man who not only invented the weather forecast but even invented the *term* weather forecast.

That date in 2017 was just a harmless bit of PR tied in with a flimsy premise, but for me it was yet another small but significant dent in the achievements and legacy of Robert FitzRoy. The shipping forecast did not begin two years after FitzRoy died. It began because he started work on it in 1854, introduced it in 1861 and worked on it so hard and zealously that it drove him to a grisly suicide.

'Sole, north-west five to seven, perhaps gale eight later, showers, good.'

The shipping forecast doesn't always go to plan. On very rare occasions it doesn't go out at all. When, in March 2010, the previous morning's forecast was read out after somebody printed out the wrong one it was the third time in three months that the same error had been made. In May 2014 the 05.20 shipping forecast didn't go out despite the fact a weather presenter was steadfastly at the microphone reading it as normal: someone, somewhere had neglected to switch over the signal from the BBC World Service, an error that went unnoticed for a good 20 minutes. There seems to have been only one instance of a shipping forecast read by two people. Tomasz Schafernaker had almost got through the 05.20 forecast on 19 December 2016, albeit with a slightly quivery voice, until he suddenly stopped, with what appeared to be a retching noise. Chris Aldridge finished the forecast for him.

'Lundy, Fastnet, north-west five to seven, occasionally gale eight for a time, thundery showers, good.'

England cricket victories in Australia have been rare things in recent years but *Test Match Special* is always there to describe them. The chances of a five-day Test finishing while *Test Match Special* handed over the long-wave frequency for the shipping forecast would appear pretty slim, but in the 2010–11 Ashes that happened three times in the five-match series, all of them won by England. The climaxes of the second test in Adelaide, the fourth in Melbourne and the fifth in Sydney all occurred during the reading of the 00.48 bulletins. *Test Match Special* continued on digital services and on BBC 5 Live, but many cricket fans were understandably a bit miffed. As the BBC pointed out, you can't move the shipping forecast on a whim. Not without six hours' advance notice.

'Irish Sea, north-west five or six, increasing seven or gale eight for a time, thundery showers, good.'

'Although Ireland's equivalent to the shipping forecast covers the entire breadth of the Irish Sea it's predominantly concerned with its coastal waters, which are defined as extending out to 30 nautical miles from the shore,' RTÉ weather presenter Karina Buckley told me. 'The geographical divisions broadcast in the sea area forecast, as we call it, aren't fixed, they vary from day to day depending on the anticipated wind conditions: sea areas with similar forecasts are grouped together in the same way as they are in the British shipping forecast.' The forecast is prepared in Met Éireann's Central Analysis and Forecast Office in the north Dublin suburb of Glasnevin by the principal duty meteorologist on shift and is issued every six hours, at 6 a.m., noon, 6 p.m. and midnight. With the exception of the 6 p.m. forecast all these are broadcast live on RTÉ Radio 1, the national public service radio station. The bulletins go out from a dedicated studio at Met Éireann headquarters and are also distributed on dedicated marine radio frequencies by the Irish coastguard together with dissemination via web and app.

'The sea area forecast begins with a description of the current meteorological situation in the vicinity of Ireland and its waters, identifying the principal features of the weather map which will have a bearing on the developing wind and weather conditions. Any gale warnings in operation are noted, together with an indication of high seas and swell, as these are not always linked to local weather conditions but can be very hazardous for seafarers. Or good news for surfers! A detailed forecast for the next 24 hours for each division of the coast follows, which begins with wind direction and speed before describing weather conditions and finally, visibility. Then comes the outlook for winds, weather and visibility over Ireland and its waters for a further 24 hours, out to 48 hours ahead.

'The sea area forecast concludes with coastal reports from five land-based stations and one of the weather buoys moored in Irish waters. This rhythmic and hypnotic list of six parameters – wind direction, wind speed, weather, visibility, pressure and pressure tendency ("falling slowly") – is the iconic sound of the sea area forecast and has even inspired an Oscar-winning song. Occasionally you'll hear my favourite coastal station name, Bloody Foreland, when the coast is carved up in the outlook. It's named after the colour of the rocks when they're hit by the rays of the setting sun, but there are quite a few gruesome myths to go with it too.'

'Shannon, Rockall, Malin, north-west six to gale eight, decreasing four or five, squally showers, moderate.'

In 1990 Stephanie Waring of Whitburn, a coastal town just north of Sunderland, gave birth to her first child. Her husband Andrew, skipper of a support vessel in the North Sea, was a regular listener to the shipping forecast and, having become attached to the litany of names helping to keep her husband safe at sea, Stephanie chose the name Shannon for their new daughter.

Seven years later Stephanie's sister, Khadine Doyle, who lived next door to the Warings, with her husband Kevin who also worked at sea, fell pregnant and decided to follow the shipping forecast theme by naming her son Bailey. She told the local press that she would have stuck with the name even if Bailey had been born a girl. A daughter followed at the turn of the millennium and was named, keeping with tradition, Tyne.

'I'm not going to have any more children and neither is Stephanie,' said Mrs Doyle afterwards. 'Besides, there are no nice shipping forecast names left.'

'Hebrides, west veering north five to seven, decreasing four at times, thundery showers, good.'

George Palin was making only his second trip to sea as a deckhand when the *Michael Griffith* steamed out of Fleetwood harbour at noon on 29 January 1953. His mother could see him at the prow of the steam trawler as it left port heading for the fishing grounds near St Kilda. He didn't turn and wave goodbye. The 58-year-old skipper, Charles Singleton, meanwhile, was suffering from a heavy cold. His wife tried to persuade him to postpone the sailing, but he assured her he was all right.

Two days later, at 09.23 on 31 January, two ships picked up a distress call: 'All ships, *Michael Griffith*, seven to eight miles south of Barra Head, full of water, no steam, am helpless, will some ship come and help us?'

It was the voice of Singleton, the only one of the 13-man crew able to work the radio. This was the last anyone saw or heard of the *Michael Griffith*.

The trawler had sailed into one of the worst storms since records began, the first casualties of one of the worst weather events in the coastal history of these islands. The combination of fearsome winds and an unusually high tide triggered a surge of water down the North Sea towards the bottleneck of the Calais Strait, causing widespread flooding on the east coast of England and in the flatlands of Belgium and the Netherlands: there were 1,836 deaths in Holland, 307 in England, 28 in Belgium and 19 in Scotland; 361 people lost their lives at sea.

Three trawlers went to the aid of the *Michael Griffith* in towering seas. The wind was at least storm force ten with hurricane-force squalls and 30-foot seas. At Orkney winds were recorded gusting up to 125mph. The 7,000-ton steamer *Clan MacQuarrie* ran aground on the Isle of Lewis, a herring fleet had gathered in the normally reliable shelter of Loch Broom near Kinross and when the seas subsided several of the ships were left in the middle of fields. The *Michael Griffith*, 125 feet long, 34 years old, didn't stand a chance. The Barra lifeboat set out into the steepling waters, later joined by its Islay colleague. As the latter was tossed around in the storm two crewmen were overcome by fumes in the engine room and died before the lifeboat could return to shore. No trace was ever found of the *Michael Griffith* and its crew – two of whom had survived previous trawler sinkings – beyond two lifebuoys that washed up on the Derry coast a fortnight later.

Further south later that day the ferry MV *Princess Victoria* sank *en route* from Stranraer to Larne with the loss of 133 lives.

Occasionally people ask what the point of the shipping forecast is these days, when we have the internet and GPS. From a practical point of view, who knows how much longer the three big glass valves keeping the BBC's long-wave transmitter at Droitwich going will last (nobody makes them anymore), but for me it's a bit like asking why we persist with broadcast weather forecasts at all. For one thing, GPS goes down. The internet goes down (recently a friend of mine who keeps a yacht at Plymouth asked me to text him the local shipping forecast precisely for that reason). Mobile phone signals can be iffy enough on land, let alone miles out at sea. In addition many smaller and older vessels plying our coastlines don't have GPS or the internet, while the broadcast shipping forecast is the only place where all the pertinent weather information for the sea areas, coastal stations and inshore waters is presented all together as one.

At any one time there are around 35,000 people at sea in the waters around our islands. Spread that figure over a year and you'd come up with a very large number indeed. Yet on average around 165 people die in the seas around the UK every year, more than half of them walkers or anglers swept from the coast. While that's still 165 people too many it's not bad considering the sheer number of people out there at the mercy of the sea and the weather. The system as it stands seems to work well, and that includes the shipping forecast. There have been hundreds of *Michael Griffithses* over the centuries, but who knows how many more have been prevented by a sensible interpretation of the shipping forecast?

'Bailey, north-west five to seven, becoming variable three or four, thundery showers, moderate.'

Ours is not the only shipping forecast, but we seem to attach more of a cultural resonance to it than other nations. Spain has one, using a map that divides the areas we know as FitzRoy and Trafalgar into smaller areas called Finisterre, Porto, San Vicente, Cádiz, Casablanca and Agadir. Their shipping forecast also stretches further into the

Atlantic with areas called Charcot, Josephine, Altair, Azores and Madeira.

In Newfoundland a shipping forecast is broadcast on the eastern transmitters of the Canadian Broadcasting Corporation, featuring sea areas whose names rival ours in terms of romance and evocativeness: Banquereau, Strait of Belle Isle, Laurentian Fan, Sable, South Labrador Sea, Funk Island Bank.

Germany's *Seewetter* forecast goes out three times a day, using many of the same sea areas as ours. Dogger, Forties, Viking and the Germanic Fischer are the same, while our German Bight is a direct translation of Deutsche Bucht. Utsira is a single area, while Südwestliche Nordsee covers Humber and Thames. Englische Kanal Ostteil is Dover and Wight, while Englische Kanal Westteil encompasses Portland and Plymouth. Other areas include Ijsselmeer, Skagerrak, Kattegat, Westliche Ostsee for the western Baltic, Belte und Sund around the Danish coast, and Boddengewesser Ost to the east of Denmark.

As for France, while they still give out the *Météo Marine* on FM it's no longer broadcast on France Inter long wave. The last forecast went out on the long-wave band on 1 January 2017, bringing to an end more than 90 years of French broadcasting tradition.

The French forecast covers many of the sea areas we're familiar with but extends right around the coast of Europe as far as Sardinia. Along the Channel Thames, Dover, Wight, Portland and Plymouth become Tamise, Pas-de-Calais, Antifer, Casquets and Ouessant respectively. The area we call Biscay is divided into four: Iroise, Yeu, Rochebonne and Cantabrico reading north to south. The area we know as FitzRoy is divided into Finisterre and Pazenne. The forecast is shared between three broadcasters, Marie-Pierre Planchon, Patricia Martin and Brigitte Forest, who are sent poems and fan letters by appreciative listeners.

'Fair Isle, Faeroes, south-west four or five, veering north five to seven, occasionally gale eight later, wintry showers, moderate or good.'

On 30 December 2013 Michael Palin was the guest editor of the *Today* programme on Radio 4. A fan of the shipping forecast, Palin asked Alan Bennett to read it during the show, selecting for him a particularly stormy one broadcast the previous October. It was quite possibly the most middle-class moment in British radio history. It's a wonder the whole of middle England didn't fold up into itself and disappear like the house at the end of *Poltergeist*. To be fair, Bennett read it straight as a die. Two years earlier John Prescott had read the forecast in aid of Comic Relief. He hammed it up. It was terrible.

'South-East Iceland, northerly five to seven, becoming variable three or four later, wintry showers, good.'

'To be honest you do want to go home at one in the morning but it is a joy, especially with the inshore waters and what have you – Cape Wrath, Carlingford Lough – you get a real geographical sense of Britain,' Corrie told me. 'You're in this little studio and you're getting storm ten and I've had the odd hurricane-force 12 over the years. I think it's a privilege to read. Not many people get to read the shipping forecast but I do. You sit here reading the 00.48 forecast completely on your own, you say goodnight, you turn out the lights, walk down through Broadcasting House and it's empty – the loneliest walk at the BBC, you can hear the ghosts of the past saying goodnight to you. Then you get in the car that takes you home after a late shift and think you're the only one up, but then people tweet at you to say thank you for the shipping forecast. They say goodnight to you.'

6

The Six Per Cent Silence: an Afternoon
with Corrie Corfield, Part Two

We moved round to the main continuity studio, 40D, a bigger, more brightly lit room where Corrie parked herself at the vast desk from which she'd spend the next few hours overseeing the smooth running of Radio 4's Saturday afternoon output. As she surrounded herself with seven screens arranged above two different mixing consoles of buttons and faders Corrie looked like someone about to produce an album for a band while at the same time keeping an eye on the CCTV covering most of central London. On the wall opposite her was a mural comprising some of the station's more familiar constructions – 'intermittent slight drizzle', 'it's five o'clock, time for *PM*', 'meanwhile, in Ambridge' – interspersed with a number of phrases that possibly only mean something to a continuity announcer: 'straight through their out time', 'are you getting my flash?' and the intriguing 'four per leg'.

With a few minutes before Corrie's first link of the afternoon she picked up the story from where we'd left off at the World Service.

'When I got married we moved to South Africa for four years because I married a correspondent and that's where he was posted,' she said. 'It was the early nineties which was probably *the* time to be in South Africa: when we arrived Mandela had just come out of prison and by the time we left he was President. I worked for a local radio station out there, which was a bit of a shock to the system after Radio 4 – having the news "sponsored by . . ." and so on – but it was great fun. When I came back my job hadn't really been filled while I'd been away so I schmoozed old Peter Donaldson a bit and haven't looked back since. It took me a long time to feel like I really belonged here, though. Peter told me it would take five years for the listeners to notice I was there, then another 15 for them to remember my name. I think it's quite right the announcers aren't personalities, but when you read the news your profile does go up a bit, especially when you're on the *Today* programme. Even then it's not like I'm John Humphrys or Sarah Montague, I'm only a tiny part of it. Announcers are the people most constantly in the listeners' lives, even though I'm just a voice being absolutely straight. It's not like I'm starting news bulletins with, "Hi folks, Corrie here," or anything like that.'

Continuity announcers are a little bit of a paradox. There's no room for them to inject much in the way of personality into their work even if they wanted to, yet their voices become a huge part of our lives. Peter Donaldson, Charlotte Green, Brian Perkins, right back through Laurie Macmillan and Alexander Moyes to Stuart Hibberd in the twenties, these are voices woven into the national fabric. When in 2017 the *Radio Times* ran a poll to find the 50 greatest radio broadcasters of all time, Corrie Corfield came in at number 23, ahead of luminaries such as John Arlott, Chris Evans and Nicholas Parsons. For someone who doesn't begin bulletins with, 'Hi folks, Corrie here,' that's an extraordinary thing.

The continuity announcer is our guide through the network and through the day. Their presence is reassuring – they're our anchor in the schedule, a voice with whom we have a curious kind of intimacy – yet on the face of it they're just the conduit between us and the station's output. 'You don't have jingles, you have us,' is how Corrie put it. It's not even as if when they read the news it's ever good: they are messengers of gloom to whom we are strangely drawn. Perhaps the mixture of warmth and authority in their voices makes them sound benevolent and trustworthy. I was half-tempted to ask Corrie if she'd record a series of bad personal news items into my Dictaphone for me to use when necessary – 'the test results are in and the prognosis isn't good', 'the cat's been run over by an estate agent in a Lexus', 'Sainsbury's are out of Frazzles' – because if I do ever have to be on the receiving end of bad news Corrie's mix of lullaby empathy and no-nonsense head girl is how I want to hear it.

They always make it sound so effortless too. The male announcers like Zeb Soanes, Chris Aldridge and Neil Sleat in particular purr into the microphone like they're in a brandy-and-cigars state of post-prandial bliss. Sometimes you can practically hear the creak of leather armchair and crackle of fire in the hearth. The continuity announcer never sounds flustered or tangibly hurried. I wonder how they manage to make it sound so effortless despite watching the clock, reading a script and sometimes having the voice of a producer in their ears or even the FM output cutting into their headphones as they reach the end of a long-wave shipping forecast.

'Broadcasting is basically telling stories,' Corrie told me. 'You're guiding people through the day or the next hour, imparting information. I like continuity to be intimate. I don't declaim, I'm more, "It's just you and me, here's the programme, now off you go." They've put a picture of the much-missed Peter Donaldson on the wall over there and I talk to him sometimes. In the old continuity studio there was a picture of dear Rory Morrison, who died awfully young a few years ago, and I'd talk to him and think, "Well, if I tell Rory everyone else will hear it too." News reading is just storytelling. You're lifting words off the page and essentially shoving them down someone's lughole. It's as simple as that, reading the news isn't brain surgery.

'I prefer a dispassionate approach and like to stay unemotional because we can tell the news is bad without the need for the newsreader to emote. When there are awful stories about awful things you don't need me to add to it by effectively saying, "God, isn't this terrible?" Radio 4 is one of the few places where you just get straight news and then a report from the correspondent. It's not two-ways, it's not opinions or speculation, it's very much, "This has happened and here's so-and-so to tell you about it." I am old school in that if I'm going into someone's house I don't want to shock them, I don't want to be an unwelcome visitor. If I make someone uncomfortable in their own home then I have failed at my job. Some people like to do witticisms after *The Archers*, or what have you, which sometimes works and sometimes doesn't. My philosophy is to just say, "It'll be back tomorrow." That's all you need.'

Like the shipping forecast, news bulletins and announcements have to be carefully timed, almost down to the second. It takes skill and experience to get it right, especially when a script can change at the last minute.

'I can look at a piece of A4, see how many paragraphs there are and be able to tell it's going to take me a minute, two minutes, or whatever,' affirmed Corrie. 'It's not so much counting words as seeing how much there is on the page. You do learn to speed up, slow down, and even edit on sight, trying at all costs not to crash the pips.'

The Greenwich Time Signal was first broadcast in the form of pips at 9.30 p.m. on 5 February 1924 at the instigation of Lord Reith and the Astronomer Royal Frank Watson Dyson. The horologist Frank Hope-Jones had suggested the idea the previous year – something for which he was so renowned that on his death in 1950 newspapers billed him as 'the six-pip man'. The sixth and final pip was extended from the standard 0.1 seconds to half a second on New Year's Eve in 1971, and in 1993 the recorder virtuoso Dr Carl Dolmetsch instigated a brief press flurry when he noticed the pitch of the pips had dropped from B natural to B flat. They're safely back to B natural now, though. Phew. The pips are, in theory, sacrosanct. There should be no talking during the pips: 'crashing' them – to be still talking once they begin – is probably the biggest *faux pas* an announcer can commit short of swearing.

'The only time people really crash the pips is when they've played a package and, for whatever reason, it's gone on a second or two too long,' said Corrie. 'Sometimes they finish with music that you can dip, sometimes they don't. There are degrees of pip-crashing: you can kiss the pips, which is when you just brush the start of them, but then there's the full-on snog and you're in there, still talking, pips pipping away at the same time. The 9 a.m. pips at the end of the

Today programme are the worst. Jim Naughtie just tried to ignore them – it was like the pips were never going to happen as long as he kept on talking.'

There are three sets of immovable time signals during the day on Radio 4: the start of the *Today* programme at 6 a.m., the opening of *The World at One* and the Big Ben chimes before the six o'clock news. Otherwise the presence or otherwise of the pips is dictated to some extent by the discretion of the announcer.

'When I'm in continuity it's my domain and I'm in charge,' said Corrie. 'Sometimes if you're coming out of a documentary or an afternoon play the drama is really tight up to the pips, in which case I won't cue the pips. If it's a difficult subject that ends in a poignant way I'll let it breathe and settle, allow for a pause and then just start in with the news. You don't have to have the pips at all costs. That way you can give the audience time to digest the last thing they've heard before you come battering in with something insensitive, as if you've not been listening.

'I'm very conscious of that. Being a continuity announcer is about being the voice of the network, but it's also about timing, sensitivity and allowing events to breathe, not being afraid to have a ten-second silence if it's appropriate. Some producer might have spent months making a heartrending documentary and you could come in and ruin the whole thing in the space of ten seconds by blundering in and not being sensitive to what's being broadcast.'

You can't really train people in that kind of thing. It's instinctive, a split-second decision in front of millions of people. It takes a real radio person to develop that knack, and few people are more steeped in the medium of radio than Corrie Corfield.

'I grew up in a house where the radio was always on,' she said. 'My grandfather was an amateur filmmaker, he won prizes, and he sold radios as a business. He was one of the first people to sell radios in Birmingham, in fact, so I grew up in a house full of them. My brother and I would make little documentaries and dramas on a tape recorder: looking back at it like that, it seems practically inevitable that I'd end up here. I distinctly remember *Listen with Mother*, so that must have been before I was old enough to go to school, and things like *Two-Way Family Favourites* because my father was in the RAF so we'd all sit down and tune in to that. *Sing Something Simple* on a Sunday evening, because it used to be before the Top 40, which you'd record trying not to catch the speech at the beginning and the end of the records.'

All the time we'd been sitting there Radio 4's output had been burbling away quietly in the background through the studio speakers. Corrie mentioned that her father would listen to *Sport on 4* and we both started singing its perky diddly-diddly-diddly-diddly-doo theme tune, at which point we both noticed there was nothing coming out of the speakers. Where seconds before Diana Speed had been giving out the news bulletin there was now silence.

Corrie bolted upright in her seat like a meerkat who'd detected a rustling in the undergrowth. 'Hello,' she said, 'what's happened here?' A glance at the clock told her the news had finished and the nation was now waiting for someone to introduce *Any Questions?* She might well have uttered a bad word at that point, I couldn't possibly say, but within seconds she'd skated her chair back in front of the microphone, tapped a couple of buttons, adjusted a fader and announced calmly, 'And now on Radio 4 it's time for *Any Questions?*'

Corrie took the whole thing in her stride but I was mortified. I'd promised not to be a distraction to her but I'd turned out to be so much of a distraction that when she should have been seamlessly guiding listeners between the news and Jonathan Dimbleby I'd managed to coax her into duetting a nostalgic *a cappella* version of Van McCoy's 'The Shuffle'.

The silence had seemed to go on for an age. When Corrie turned back to me my eyebrows were somewhere high on my scalp and my mouth was opening and closing like a carp that had just heard surprising news.

'That's all fine,' said Corrie cheerily as she pushed buttons again, despite being faced with somebody who by all appearances seemed to be having a stroke. 'I should have played a trail but we'll have that afterwards instead. There was only a nice little pause. Thank goodness I'd loaded *Any Questions?*' Realising that she clearly wasn't going to get much sense out of me for a while, Corrie continued, 'We no longer have a studio manager, which is a big change. If we still had them, just then a studio manager would have given me a 30-second warning before the end of the news. Instead, people probably just think I'm drunk under the desk or have been locked in the loo or something.' She checked Twitter: only one person had commented about the brief pause, which seemed a pretty safe barometer reading.

'The first big time horror I had was in the late eighties when our continuity used to be in a different part of the building, and they were doing some building work, and they cut through the major electrical cable and absolutely everything went out. By instinct I went to the microphone to apologise and let everyone know, but of course it wasn't going anywhere, the microphone was as dead as everything else. Radio 4 has this emergency digital recording; there's one in the control room here and there's another at the transmitter in Droitwich. If there's a silence of two minutes it cuts in automatically, playing a holding message until we can sort out what's going on. We have to remember to override it before the two minutes' silence on Armistice Day, and, of course, we have three minutes before the nuclear submarines start to worry.'

At that, what little blood remained in my face drained away with the sound of a child sucking the dregs of a milkshake through a straw. If the nuclear submarines hear Radio 4 has gone off the air for three minutes they start tapping in the codes to fire nukes at the bad guys. I'd thought it was just an urban myth. The silence I'd caused had, I estimated, lasted a good ten seconds, meaning that I had personally just brought Britain six per cent closer to nuclear Armageddon than it had been probably at any point since the end of the Cold War.

'We also have pre-recorded breakdown passages now, so if there is a major problem I can do what needs doing rather than just apologising and playing music and reading bits out of the *Radio Times*,' Corrie continued. 'And we do have a bank of programmes to put out if it's a serious problem that will take time to fix. I still have anxiety dreams of running to the studio knowing I'm not going to make it in time, or finding the door locked. I've been quite lucky, but there have been times when I've played the wrong programme, or the line to a correspondent has gone down. If there's a major breakdown there are things you can play to keep people's radios burbling away while you're running around trying to get to the bottom of the problem.

'I think the biggest change in my time is that as announcers and newsreaders we sound more relaxed than we used to. If you listen back 30 years we sound of a different era, a different time altogether. I don't think it's been done deliberately, and it's certainly not dumbing down, but a range of accents has become more acceptable and received pronunciation has largely gone, yet there is still a definitive Radio 4 sound. I sound very different from Harriet Cass and Charlotte Green. Peter Donaldson had a very precise, clipped way but even he, at the end, was different to how he sounded in the seventies. I think it's just been a general process of osmosis. The correspondents are so different now. Listen to Richard Dimbleby talking about the liberation of Belsen and it's all very portentous but correspondents now don't sound like that, they're really in the moment, and I think that rubs off on us. We still have very high standards in terms of being grammatically correct, and we use vocabulary you wouldn't necessarily hear on other networks, but I think we've lost a bit of the aloofness Radio 4 used to have. It's just reflecting life, I think, responding to cultural changes in people's listening habits, social media and that kind of thing.'

There are, Corrie informed me after a rough count in her head, ten or 11 staff announcers on the network, not all of them full time, while there are also roughly the same number of freelancers on the books too. There are broadly three shifts: day continuity, from 8 a.m. to 5 p.m.; mid-continuity, 11 a.m. to 8 p.m.; and late continuity, 4.30 p.m. to 1.30 a.m.

'The other side of the job is reading the news,' she said. 'But not everyone does that yet – although I do think in future it will become more interchangeable. We also have to read the news on Radio 3 at 1 p.m., 5 p.m. and 6 p.m. too these days.'

·The announcers have some input into the links between programmes: 'Every programme comes with a Presentation Detail, which may have an intro and an outro that includes presenter and producer info and so on. What we do is use that in the mix with points ahead, trails, mentions of other programmes, etc, which we do write ourselves and pull together to make everything seamless.

'Everyone has their little quirks. I have a real problem saying the word "tuberculosis", so if it comes up in a news bulletin, or I have to introduce a documentary about it, I just call it "TB". I have another colleague who can't say "phenomenon". I find "judiciary" and "judicial" quite difficult for some reason too. On the *Today* programme once I said "irreVOKEable" instead of "irREVocable" and I knew as soon as I said it it was wrong, but at the time I couldn't work out why. Someone had a go at me on Twitter about that one, and rightly so.

'I still get nervous sometimes, yes,' she admitted. 'Usually if I'm a bit tired or if there are other people in the room. At one point I developed a real thing about the 7.30 a.m. news summary on the *Today* programme. I don't know what it was but I used to practically have a panic attack and was convinced people could hear that I was struggling. I've done the *Today* programme for years and have read countless news bulletins sitting next to John Humphrys, so I don't know why it became a thing, but I thought, "I'm going to have to take charge of this because people will hear that I'm nervous and I can't have that." It's a two-minute bulletin but I made them write it in the script as one minute and 50 seconds, psychologically keeping it under two minutes, and once I started doing that it was fine again. Very strange. It does happen, though. It usually happens when a story is added in at the last minute and you've not had a chance to read through it first. Once I was reading off the computer screen during *PM* and someone downstairs in the newsroom chose that moment to alter something in the script and the whole thing disappeared from the screen. I just had to say, "I'm sorry, I can't carry on."'

At the end of the afternoon, as I packed away my Dictaphone, I reflected that although it was the first time I'd met Corrie it felt as though it was the extension of a conversation we'd been having for years. Corrie's voice has been a constant presence in my life since my teens – longer than some of my best friends, less troublesome than some of my relatives, more enduring than most of my relationships. Being an announcer is an unsung position – and I got the impression they like it that way – yet these voices have marked out our lives.

'Peter Donaldson used to say, "The minute you don't get that slight rush, the awareness, that sense of 'goodness me this is live', that's when you don't want to do it any more,"' said Corrie as we emerged into reception again. 'It's a performance. Even actors who don't get stage fright, I bet they still get a little *frisson* before the curtain goes up.'

7

'A Science of Inestimable Value to Humanity': Guglielmo Marconi in Dorset

Sandbanks is an odd kind of a place. Part spit of land, part peninsula, it's a two-mile long skinny mini-Portland suspended from the Dorset coast at the mouth of Poole harbour. It is also eye-wateringly wealthy. There are probably houses behind all the security gates you pass, but you can't see most of them. In fact I know there are houses behind the security gates because every now and then one is put up for sale and the newspapers go absolutely bananas.

Before I'd set out for Dorset I'd typed 'Sandbanks' into the search box of a property sales aggregation website, listing the most expensive places at the top. Scrolling to the foot of the first page I was still looking at a shade under £3,000,000. Three. Million. Pounds. Paris, Manhattan and Singapore are cheaper places to buy property than Sandbanks. Imagine that. 'Shall we move to the outskirts of Bournemouth?' 'Oof, don't know about that, Lower Manhattan's more in our price range.'

I felt awfully scruffy driving along the spit. For a start, I was driving rather than arriving by helicopter. Also, I don't own a single pair of deck shoes. Or a yacht. My car looks at the best of times like it's been recently rolled in a peat bog, let alone next to the giant gleaming black-windowed behemoths gliding along the same road looking like the result of a limousine being rodgered by an armoured personnel carrier. I could almost sense house prices tumbling as I approached before returning to their whopping equilibrium after I'd passed.

The Haven Hotel is right at the end of the Sandbanks spit. It lives up to its name, too. I'd spent the last couple of miles of the journey convinced that at any moment some kind of riff-raff police would step into the road with a hand raised and turn me away; that my sort wasn't welcome round here and I'd be better off heading back towards Poole sharpish where – and here they'd look me up and down – there were hairdressers, car washes and launderettes. Pulling into the car park I was overwhelmed by a sense of relief as I switched off the engine. The Haven Hotel looked quite normal.

In the global property development wars Sandbanks is ground zero. You sense that if someone even mentions the possibility of maybe thinking about selling their house while standing in their kitchen with a cup of tea, within seconds property developers are staging a fight to the death on their doorstep for the chance to knock the house down and build a hive of luxury apartments in its place. The Haven Hotel is a welcome antithesis to this Wrecking Ball of Damocles, at least for now, helping Sandbanks retain some of its character and historical context in the face of the relentless onslaught of the luxury apartment.

The hotel's pavilion design harks back to a different age. There's a classy dignity about it. It's well kept, the white paintwork on the outside immaculate even when I visited in the off-season. Up close there are a few minor signs of age but among the dull parade of back-of-a-fag-packet modern architecture that surrounds it the Haven Hotel hints at history.

My room was at the far end of a long corridor, one of those undulating corridors with creaking boards that are the sign of an agreeably venerable hotel. It was midweek when I visited and almost midway between the end of one summer season and the beginning of the next, the low-turnover solstice of the hospitality trade. The hotel was quiet and I could tell it was a good couple of months since the corridor had been busy: the air I passed through felt like it hadn't been seriously disturbed in weeks. This was a corridor where ghosts from the golden age of the British seaside could winter undisturbed.

I'd pulled in at the last vestige of dusk, and by the time I'd checked in and negotiated the corridor darkness had fallen. I looked out of the window at the clear night sky and thought about how, more than a century before, it would have been lit up with flashes from the experiments that took place on this site. For by staying in the Haven Hotel I was spending the night in a cradle of radio history: Guglielmo Marconi kept a base here throughout the key development years in the story of radio, arriving in 1898 and leaving in 1926 when the old building was knocked down and the current one put up in its place.

There's a slightly battered plaque on a slightly battered pillar outside the hotel, a sentinel of what must have once been the main entrance but is now off to the side, almost forgotten, close to the little quay where the chain ferry connects Sandbanks with Studland on the other side of the harbour. I went out to look for it the next morning and even though I knew what I was looking for it took a little bit of finding, but it's there all the same, cracked and faded:

GUGLIELMO MARCONI
(1874–1937)
Inventor of a
Wireless telegraphy system.
He conducted experiments from
Sandbanks over a
30-year period.

That does understate things rather.

Marconi settled in Britain in 1896, turning up at the house of his uncle in Bayswater with his mother in tow. He was 21 years old and already frustrated in his ambitions for the world of electronic communication. He came from wealthy stock, his father a silk merchant and his mother the granddaughter of John Jameson, founder of the Irish whiskey distillery. He studied under Augusto Righi, a Bolognese pioneer in the field of electromagnetism who some claim was the first to generate microwaves, at the University of Bologna but had to do so unofficially as he'd failed to gain admission as a regular student. Guglielmo Marconi, the man who would do more than most to turn brain-busting physics theory into a commercial reality that would change the world, was not a natural scholar.

With Righi's help the young Marconi expanded on the work done by radio pioneers like Heinrich Hertz, Édouard Branly, Nikola Tesla and David Edward Hughes, taking the results of experiments and discoveries that had never left the laboratory and turning them into a plausible and commercial method of communication. In his attic Marconi worked with elevated aerials consisting of copper plates suspended from bamboo frames and used a coherer – a glass tube that could detect radio signals thanks to some jiggery-pokery involving two electrodes and a scattering of iron filings contained inside – and a battery to make a bell ring up in the eaves of the family villa. Unlike similar figures in laboratories across Europe and America, however, Marconi didn't view wireless technology as a mere conversation piece for eggheads. From his earliest experiments he recognised its importance to the future of ship-to-shore communication – and also the potential commercial value of wireless telegraphy.

In 1895 he was able to move his apparatus outdoors in order to test over longer distances, but was at first frustrated when it appeared the signals he was sending could only travel for around half a mile, the distance British physicist Sir Oliver Lodge had declared to be the maximum possible range. Marconi was convinced his wireless signals could travel further, however, and raised the height of his antenna. One early breakthrough he made was in realising that radio waves did not travel in a beam, like light, but spread over a wide area, meaning the receiver did not have to be in the direct line of sight of the transmitter. Marconi had been sending his faithful valet, a man named Mignani, out across the family estate and up a hill with a receiver, where he would wave a white handkerchief whenever one of Marconi's transmissions hit the spot. One day in the spring of 1895 he sent Mignani over the brow of the hill until he was entirely out of sight, this time with a souped-up telephone receiver in one hand and a revolver in the other that he was to fire when he received a transmission from the boss. Mignani marched off into the distance looking not unlike a man about to commit an elaborate suicide, crested the hill and disappeared out of sight. Marconi stood at the window, watched Mignani drop behind the hill, waited a couple of minutes until he thought the valet was ready, then tapped out a message. A few seconds passed during which Marconi could hear his heart thumping in his ears. A crack of gunfire, a scatter of birds from the trees, and Mignani appeared on top the hill again, handkerchief aloft.

Later that year, when he'd written up his findings and was sure of their commercial potential, Marconi approached the Ministry of Posts and Telegraphs who, to his horror, showed not the slightest interest in investing in or acquiring his work. The young entrepreneur was convinced of the value of his findings to the maritime industry, and was sure the Italian authorities would see it that way too, but as he gushed excitedly about the potential of wireless telegraphy the men from the ministry pursed their lips, twiddled their pencils, looked out of the window and said they couldn't see the advantage of this kind of oompus-boompus when they already had cable telegraphy. To Marconi this was a bit like asking why anyone would need a personal mp3 player when they could just push a record player around in a shopping trolley, but the bigwigs were not to be convinced. Marconi packed up his diagrams, his apparatus and his commercial nous and headed for London, where he knew there were men like Hughes and Oliver Lodge who might lend a more sympathetic ear.

Having arrived in February 1896, by early June Marconi had already filed a patent for 'improvements in transmitting electrical impulses and signals and in apparatus therefore'. He was still only 22 years old and would become a regular face at the patent office: every time he made some kind of breakthrough or improvement on his existing work he'd hotfoot it to their door to take out another patent, hoping to ensure he dominate any future market for wireless telegraphy. This was no simple act of administration either: Marconi would become notoriously litigious over the coming decades, firing off writs if he felt anyone, even in all innocence, was looking to tread on his handmade brogues.

When he arrived in Britain Marconi soon acquired a letter of introduction to the Welsh physicist William H. Preece, a leading light in British wireless research who'd risen to become Engineer-in-Chief at the GPO, overseeing the development of Britain's early telephone networks. Preece had been instrumental in improving the signalling systems of British railways and, in 1889, was part of a group who successfully sent a Morse code message over a mile across Coniston Water in the Lake District.

The ageing Preece took to Marconi immediately, drawn to his youthful vigour and insatiable drive. Within days of presenting his letter of introduction Marconi had set up a demonstration of his work, using a spark transmitter (a device that creates electromagnetic waves by sending a spark across a gap between electrodes) to send London's first wireless telegraph from the GPO headquarters in St Martin's-le-Grand to another GPO building half a mile away on the Embankment.

Impressed, Preece granted Marconi access to his personal laboratory and even assigned him staff. Marconi's subsequent progress was swift. Within a few weeks of that first meeting he'd managed to send a wireless telegraph over two miles of Salisbury Plain, and the following summer, assisted by George Kemp, a GPO man assigned by Preece who would go on to be one of Marconi's most loyal and trusted confidants, he extended the transmission distance to three miles by sending a message from Lavernock Point to the island of Flat Holm in the Bristol Channel. Within a week of that, a transmission from Lavernock Point was received nearly ten miles away at Brean Down in Somerset. Noting how effective the experiments had been over water – at this stage the commercial opportunities of ship-to-shore communication was all he was interested in – Marconi set up a research station at the Needles on the Isle of Wight and, still only 22 years old and invigorated by the rapid progress he was making, heeded the advice of his uncle Henry Jameson Davis and set up a private company rather than bring the government in on the scheme. On 20 July 1897 the Wireless and Telegraph Signal Company was formed.

In 1898 Marconi sized up the Sandbanks peninsula as another potential base for his operations. It had the advantage of being surrounded by water yet was accessible along a causeway. At 16 miles it was a good distance from his Needles base, and although tourism in the area was just starting to pick up, Sandbanks remained relatively undeveloped and free of hordes of people. The Haven Hotel stood almost alone on the southern stretch of the peninsula and had been recently acquired by a French couple named Poulain, with whom the Italian built an immediate rapport. While other Marconi stations would come and go he kept a base of operations at the Haven Hotel for nearly 28 years, sometimes living there himself for weeks at a time. His yacht the *Elettra*, meaning 'spark', was often seen moored at Brownsea Island, a short hop across the harbour from the Haven, and he even spent his honeymoon there. Marconi was never really a sentimental man but something about the Haven Hotel and the harbour it surveyed got under his skin: he was known to take the train from Waterloo after a day working in London purely in order to dine at the Poulains' estimable hotel restaurant.

I ate in its modern equivalent, the huge dining room empty save for two or three tables, and looked out into the darkness of the harbour where red and green lights flashed on the buoys that marked the shipping lane. A trawler chugged past, floodlights on the wheelhouse pointing the way and lighting up the patch of water at its bow in silvery shards, the low, throaty rumble of its engine faintly tangible through the floor.

After dinner I adjourned to the Marconi Lounge, empty of diners but inviting, all dark beams and a fire crackling away in the grate. Delightfully understated, the internal lounge is surrounded on all sides by the restaurant, bar, reception and kitchens, and contains a couple of framed photographs of the man himself as well as an old French poster advertising a brand of Marconi radio. On one wall, out of the way and almost in a corner, was fixed a small plaque that looked as though it had predated the current décor by a considerable margin. 'In this room that may truly be called the cradle of wireless,' it began, confidently in faded gold lettering on a black background,

'Guglielmo Marconi, during the years 1898 until 1926, conducted some of his most important experiments in wireless telegraphy and telephony and laid the secure foundations of a science of inestimable value to humanity.'

It certainly knocked a few spots off 'he conducted experiments from Sandbanks over a 30-year period' anyway. Whether the room could truly be called the 'cradle of wireless' is obviously up for debate. Its sofas were comfy, its fire crackly, its celebration of Marconi tastefully low key. The hotel was modified extensively in 1926, so there's a chance the room isn't exactly as it was in Marconi's day, but as a quiet area to sit and look into the fire it feels like a fitting way to commemorate the significance of the location. The contrast with the excitement of the experiments that were carried out here was marked, and the peace and quiet made the history more real, somehow. In the low roar of the flames it was almost possible to hear the faint sound of Marconi accompanying himself on the piano, entertaining staff and friends with songs after an agreeable meal gathered around a long communal table in the restaurant.

Marconi spent happy times at the Haven Hotel. It was an escape as well as a place of work. It was also a constant in his progress from excitable young inventor hauling his demonstration devices between sceptical eminent scientists to one of the most successful business people in the world. He brought his family to the hotel, entertained friends, worked and came as close as he ever did to relaxing within its walls. Looking out through the restaurant to the darkness of the harbour beyond, it was possible to imagine how peaceful it must have been at the dawn of the twentieth century: a true haven indeed, one that provided emotional equilibrium for the man who did more than most to make the world a smaller place by connecting people, nations and continents.

Yet today, with Sandbanks being Sandbanks, the days might be numbered for this quiet, dignified piece of radio history stuck out at the far end of a topographical anomaly in the far south of Britain. For all its charm, for all its understated grace and elegance, and the warmth of the welcome you receive at the Haven, there's more profit to be made from the land on which it sits. This symbol of what made Sandbanks special in the first place is earmarked for demolition by a property developer on the grounds that 'if you went to Palm Beach you wouldn't expect to find hotels there that are a hundred years old'. The Haven and a couple of other hotels closer to the mainland have, apparently, reached the end of their 'economic lifecycle', whatever that means, and plans have been drawn up for a ten-storey complex of nearly 200 luxury apartments. One report even mentioned a proposed huge rooftop restaurant called – and I'm not making this up – Jurassic Eye. The plans have received thousands of objections, in what certain sections of the media are calling the 'Battle of Bling Beach', but money talks, and on the Sandbanks peninsula its voice is foghorn-loud.

While Marconi spent nearly three decades of happy and productive times in Dorset he was, after all, a man with a keen eye for the main chance. Would he have been lying down in front of the bulldozers? Probably not. He might even have slapped down a deposit on one of the penthouses, but that doesn't make the removal of the last link to Sandbanks' place in the heritage of world radio or the visual focus of the peninsula from the rest of the harbour any easier to swallow.

At breakfast the next day it was such a beautiful morning I couldn't fail to be in better form. The sun rose milky yellow over a blue sea and arrived in perfect time for high tide. A fishing boat cruised past the window at the prow of widening wake surrounded by two dozen swooping seagulls. A short time later a little pilot boat motored the same way followed by an enormous ferry heading for Cherbourg that pitched the dining room into sudden shadow, passing so close that if the window had been open I could have reached through it and rapped on a porthole. Gradually the ship turned towards the horizon and became a hazy grey silhouette. If ever there was a time and a place to imagine the possibilities that lay over that horizon a sunny morning at high tide from the dining room of the Haven Hotel would have been it.

As the twentieth century dawned Marconi widened his ambitions from linking mere nations to linking entire continents. In the autumn of 1900 he built a wireless station at Poldhu on the coast of the Lizard Peninsula in Cornwall, and towards the end of 1901 he took a liner across the Atlantic to Newfoundland, setting up a temporary wireless station in a disused, wind-blasted diphtheria hospital on the appropriately named Signal Hill overlooking the city of St John's on Newfoundland's eastern tip. This was his biggest test so far and it would be a very public one at that: scientific consensus was that radio waves travelled in broadly straight lines in the same way light did. While Marconi's hilltop experiment had been a success, what he had demonstrated was basic diffraction. Many scientists believed the curvature of the Earth made it impossible to send wireless signals between Europe and North America because somewhere over the Atlantic they'd shoot straight out into space. In fact the ionosphere, an electrically charged atmospheric layer around 50 miles above the Earth, bounces and bends radio waves back toward the surface, and this is where the success or otherwise of Marconi's Newfoundland experiment lay in probably the biggest gamble of his career. His furthest successful transmission to date had been received around 80 miles away – and from 80 to a couple of thousand was quite the jump. In effect Marconi was attempting to leap a dozen double-decker buses on his bike having only just mastered the wheelie.

Thursday 12 December 1901 dawned cold and blustery in Newfoundland, and the Marconi employees, marshalled as ever by Guglielmo's faithful assistant George Kemp, were having trouble sending up the kites from which the antennae would be suspended ready to harvest from the ether the transmissions from Poldhu. The whole project had seemed doomed from the start: the original masts at Poldhu had blown down in a gale, as had the apparatus at their first choice for a receiving station on the western side of the Atlantic, Cape Cod. Having headed north to Newfoundland instead they were beset with problems raising the antennae – at one point they'd seriously considered nailing one to an iceberg – and even once they were up Marconi and Kemp spent three whole days and nights at the receiver waiting for a signal.

Eventually, at about 12.30 on the afternoon of 12 December, Marconi's eyes suddenly widened and he pressed the telephone receiver to his ear. He was sure he could hear successive bursts of three dots – the Morse letter 'S' – sent as arranged from Poldhu some 2,200 miles away. He handed the earpiece to Kemp to confirm what he thought he'd heard. Kemp listened for a while and confirmed that yes, he could hear a three-dot signal too.

Marconi's trust in himself, his equipment and his theories had, it seemed, been justified. In subsequent years doubts have been expressed that a signal sent in daytime on a medium-wave frequency could possibly have travelled that far. What Marconi had thought was a three-dot Morse code 'S', doubters surmise, was interference or even a harmonic echo of the signal rather than the signal itself. They had, after all, spent the previous three days listening at around the appointed daily time arranged with Poldhu to the static of the atmosphere, with all its pops, hisses and squeaks, not to mention the constant noise of the wind outside: you'd maybe forgive the odd aural hallucination in those circumstances.

Whatever the truth behind the Poldhu-to-Signal Hill test transmission the technology worked: it was only a little over a year before regular wireless signals were being sent to and fro across the Atlantic. In 1903 the US President Theodore Roosevelt sent a message of greeting to King Edward VII in London and four years later a regular transatlantic wireless service was introduced between the Marconi station at Clifden in County Galway, Ireland, and Glace Bay in Nova Scotia.

In 1909 Marconi was awarded the Nobel Prize for Physics, the first time it had been awarded for the practical application of science rather than to a theorist. Curiously the committee announced the prize as shared between Marconi and Karl Ferdinand Braun, a German scientist who, while he had made some progress in the science of wireless telegraphy, wasn't previously regarded as a serious contender. In addition he and Marconi had also been involved in a heated dispute over a patent (but, then again, Marconi was involved in heated disputes over patents with just about everyone, so we can probably let that one pass). Marconi later said that he'd found himself sharing a train compartment with Braun on the way to the ceremony in Stockholm and that Braun had offered to play cards for the whole prize fund. The Italian declined on the grounds he didn't consider himself a good enough card player.

The committee had considered giving the prize to Orville and Wilbur Wright, who had continued their pioneering work in aviation since their first flight at Kitty Hawk, North Carolina, in 1903, and looked at a number of European-based aviators but, with this flying business looking a bit dangerous, they feared a public backlash if any airborne Nobel laureates were subsequently killed in a tangle of canvas and balsa wood after losing a toe-to-toe with gravity. 'In the present state,' noted the committee, 'this invention can hardly be considered to benefit mankind.'

Marconi continued to go from strength to strength in the years leading up to the First World War. When the *Titanic* launched in 1912 its radio operators were Marconi employees (the man himself had been offered a free passage on the ship but elected to travel on the *Lusitania* instead, departing three days earlier, apparently because he had a great deal of work to do on the voyage and liked the *Lusitania* stenographer better) and it was the distress messages broadcast from the telegraph room of the ailing liner that eventually summoned the *Carpathia* to pick up the 700 survivors and deliver them to New York, where Marconi was waiting to talk to his lone surviving wireless man.

Marconi's world involved being at the centre of the news in some of the globe's most glamorous locations, but also in remote, godforsaken places where he set up his wireless stations. Somewhere in between was Chelmsford, where he established the headquarters of his business in a former silk works in 1898 and then, in 1912, built the world's first purpose-built radio factory at New Street in the town.

Yet even Chelmsford is a throbbing metropolis compared to one of Marconi's more out-of-the-way stations on the wind-blasted, Atlantic-lashed west coast of Ireland. Sitting looking out at Poole harbour in the sunshine, some idle daydreaming about Marconi had taken me back to the bog that conquered the world.

A few minutes out of Clifden, County Galway, I turned off the road onto a single-lane track. A few hundred yards ahead a steel-barred gate blocked the way, from which hung a wooden sign with 'NO QUADS' sloshed on it in red paint. I parked the car just off the track, went through the gate and set off on foot heading for a bog that looked like any other in the wilds of Connemara but was actually of rather extraordinary renown.

Some ridiculous things are held up as having 'changed the world' these days, but this one genuinely deserves the epithet. It may have seemed like the end of the world, a geologically turbulent, inhospitable backwater, but this rocky, craggy, boggy, apparently unremarkable patch of Connemara where black-faced sheep roam free and quads are not welcome had hosted both the first commercial transatlantic wireless communication and, barely 500 yards away, the landing of the first ever non-stop flight across the Atlantic.

After a walk of ten minutes or so I rounded a drumlin and saw a giant white-painted stone monument in the shape of a cairn. Nearby, about 50 yards away, some sheep mooched half-heartedly about on a flat patch of old concrete that looked to be the foundations of an old building. Grass grew through the cracks and moss speckled the remains, at the heart of which were what looked like two enormous rusted metal fastenings sticking out of the ground – massive dark orange things, the shape of Liberty's torch. I didn't quite fall to my knees like Charlton Heston at the end of *Planet of the Apes* but I did experience a *frisson* of realisation that I'd reached the spot I was looking for.

Marconi's scouts had found this location not long after the dawn of the twentieth century. Half-Irish himself and married to the daughter of Lord Inchiquin, Marconi had long-known that the west of Ireland was probably the best place for him to construct a wireless station capable of launching and receiving messages between Europe and North America. The Derrygimlagh bog was perfect: close enough to the coast to catch the signals at the earliest moment, remote enough that the work could be done free of disturbance. There was even fresh water and turf close at hand to keep the complex more than adequately supplied.

When construction had finished in the autumn of 1907 the site would have resembled the most fantastical thing in the world. There were eight wooden masts more than 200 feet high connected by wires that stretched for half a mile across the bog. The boiler house had six tall chimneys that belched smoke constantly, and there was even a narrow gauge railway on which three steam locomotives puffed back and forth. Had I stood here a century earlier I would have been at the centre of an industrious bustle of activity at the very cutting edge of technology: men cutting away at the turf to stoke generators, the small railway line ferrying people to and fro and pulling cartloads of coal, and a constant stream of engineers, scientists and clerical staff travelling up and down the road I'd just trundled up feeling like the only person in the world. Back then the place was practically a town in itself, and in the dark for miles around people could see strange flashes in the sky and hear cracks and booms that broke open the night, all emanating from the mysterious complex built at the heart of the bog.

We think nothing now of instant communication. We take it for granted that we can WhatsApp someone in New York and receive a reply within seconds on our tiny, portable handheld devices (or get all haughty if a reply doesn't come back within five minutes of sending the question). That we can do this is thanks in large part to Marconi and the very wireless messages that would emanate from and terminate on the bog where I stood being regarded with increasing suspicion by some sheep.

The giant building that housed the condenser by which sound waves were converted into electrical signals contained 1,800 sheets of iron, each measuring 30 feet by 12, suspended a foot apart from the ceiling. This was where the electrical charge needed to power the messages was formed, and to earth them were two sheets of copper mesh, each of which was 600 feet long. It must have been an extraordinary sight, the seemingly magical properties of which were mooted by a local newspaper who blamed the week of torrential rain that followed the plant's opening on the clouds having been penetrated by electricity generated at the Marconi station.

The site also boasted a blacksmith's forge, vegetable garden, bungalows capable of accommodating eight families, a grocery shop, a recreation centre that featured a reading room, billiard room and even an outdoor tennis court. Connemara had never seen anything like it.

After a short period of testing the first official messages were sent across to Glace Bay, Nova Scotia, messages that encapsulated perfectly the prevailing mood of the time. One relayed British Prime Minister David Lloyd George's congratulations to Marconi, who was over at the Nova Scotia station at the time, on behalf of the British Empire and its dominions, while the other was from a local county councillor to Theodore Roosevelt, hoping that he could count on the president's backing for Irish self-government.

The station proved to be a great success, not only in terms of technological advancement but in aiding the local economy: around 300 people were employed at the station at its busiest periods. On a windy grey day picking my way over a few concrete blocks and giant rusting lumps of metal in the grass it was hard to imagine the sheer industry that had gone on there a century earlier. Nature was slowly reclaiming the site: the old route of the railway tracks could still just about be seen and grass was growing up between the cracks in the concrete, moss blurring the straight edges between slabs. A couple of the sheep came to join me, looking faintly confused by the whole business, and eventually followed me over to the giant white obelisk I'd passed on my way to the old Marconi site.

For this patch of bog to have witnessed one world-changing event was remarkable enough, but the white conical monument a stone's throw from the old concrete slab commemorated another. A small plaque on the cairn pointed an arrow southwards indicating a spot around 500 metres away where at around half past eight on the morning of Sunday 15 June 1919 a faint, distant and unfamiliar noise could be heard that would turn that day into one like no other. The noise got louder until it became the throaty roar of an engine and, at the peak of the crescendo, a flimsy-looking twin-engined biplane appeared out of the murk. It circled the Marconi station a couple of times then flew off towards Clifden, which it also circled. Most of the townspeople were at Sunday mass so only heard the engines; eventually the noise that had caused glances to be exchanged in the church as the priest intoned solemnly faded to nothing. Back at Derrygimlagh the plane appeared again, circled, banked and descended, coming to earth with a bump, a roll, another bump and then eventually a small crash that left the machine in a frankly undignified position with its nose buried in the mud and its backside in the air.

A couple of telegraph operators on duty at the Marconi station ran towards the aircraft, from which they saw two men hop out into the mud and start walking briskly towards them. When they met, the bigger of the men said, 'I'm Alcock, just come from Newfoundland.' His colleague, a Lieutenant Brown, added, 'Now *that* is the way to fly the Atlantic.'

The exhausted-looking plane that lay face down in the mud had over the previous 16 and a half hours flown John Alcock and Arthur Brown into the history books as the first men ever to fly non-stop across the Atlantic.

The Marconi station had been garrisoned by the British Army during the First World War, when the troops had been careful to prevent the station being used to broadcast any mention of the 1916 Easter Rising to the US. During the War of Independence, the notorious Black and Tans used the social facilities, and after partition the Republicans considered it a Free Staters' stronghold. Those associations are what sealed its fate, and one night in 1922 the place was attacked, burned and all but destroyed, which is largely why today there is nothing left to see of this pioneering centre of communications save for some concrete slabs and a couple of lumps of rusting metal.

In the meantime Marconi's equipment and technology was improving all the time, and the need to be so close to the Atlantic coast rescinded. Even before the attack by the Republicans Marconi had been persuaded not to close the plant itself (at the loss of hundreds of local jobs), but by the time of its destruction his main operations had moved to Wales. He divorced his Irish wife in 1927 and returned to Italy, where he embraced Fascism with vigorous enthusiasm and became close to Benito Mussolini, who on Marconi's death in 1937 took it upon himself to design the mausoleum in the grounds of Marconi's home in which his remains rest to this day. At Marconi's funeral the largest wreath was sent by Adolf Hitler. Despite this ignominious sunset to his remarkable life, Marconi's name still remains fêted in telecommunications. While not as thrusting a pioneer as many in the same field, the fact remains that what was achieved in Sandbanks, Nova Scotia, the Isle of Wight and in that remarkable patch of bog was utterly extraordinary.

I hope the Haven Hotel is spared the whims of rapacious property development. Tangible British radio heritage is sparse enough as it is, and there is so much history infused into the place that it deserves a reprieve. But, then again, why indulge the past of a medium that people don't have to pay for? The bottom line is, after all, the bottom line.

8

'The World is Calling for More':
the Chelmsford Broadcast

Britain woke on 15 June 1920 to news of industrial unrest. This wouldn't have come as a great surprise: in the immediate postwar years Britain woke pretty much every morning to news of industrial unrest, but on that day in particular from farm labourers to Welsh miners, Lever Brothers' soap workers in Liverpool to boat builders on the Thames, unhappy workers were either striking or on the verge of striking. At midday hundreds of members of the Association of Wireless Telegraphists employed by Marconi walked out, frustrated that their demands for a minimum wage of £2 16 shillings for men working 12-hour shifts, seven days a week, were not being met by the company.

Marconi's Chelmsford headquarters was all of a tizz that morning, but it wasn't with talk of striking marine telegraphists. Instead the worried faces, tugged bottom lips and furiously polished spectacles betrayed the nervous anxiety throughout the complex ahead of an event that evening that could, if all went smoothly, revolutionise the fledgling technology of wireless telephony.

All around the country enthusiasts were willing the evening to come so they could gather round their sets. Office clocks, school clocks, workshop clocks, pocket watches: timepieces everywhere were stared at and glared at like never before. Long before the end of the office day desks were tidied, coats were draped over forearms and hat brims were fiddled with. Dinners were taken early, at least by those whose excitement hadn't extinguished their appetites altogether.

In North Lane, Canterbury, early that evening a gathering assembled at the premises of S.W. Bligh, electrical goods retailer, including Count Louis Zborowski, the racing driver who would become the inspiration for *Chitty Chitty Bang Bang*. In Oxfordshire, electrical engineer and keen wireless enthusiast Hubert Treadwell showed a group of friends into the receiving room he'd set up in the old cottage he used as a workshop next to his home in Middleton Cheney where, at 5 o'clock, he had already picked up a test broadcast of a man singing 'Land of Hope and Glory'.

The Derby Wireless Club, believed to be the first of its kind in Britain having been founded as far back as 1911, prepared to receive members and their guests at the old Bank Chambers on Iron Gate. Its founders, Sancroft Taylor and Alan Trevelyan Lee, had been among the earliest adopters of the new technology, sending messages between their houses on homemade equipment before Taylor went on to serve as a wireless operator during the war.

Geoffrey Gore, manager of Chase Motors Ltd of Berwick-upon-Tweed, had also been a wireless operator during the conflict and was looking to expand the business into that area by the adopting the slogan 'the wireless specialist of the borderland'. That night's event would, he hoped as he straightened chairs and awaited his invited guests, cement that reputation. If one sourced the required components themselves, he'd explain, the only significant costs in building a wireless set were the valves and headphones, items he could provide. A self-built set was not out of the financial reach of ordinary folk, he'd tell them, with his own, the one they were about to hear, having cost less than £5.

Farther down the east coast at the Alexandra Theatre in Kingston upon Hull, musical director Charles Greave prepared his wireless receiving set. The Imperial War Museum's RAF section, housed at the Crystal Palace, opened up its galleried entrance hall to visitors who arrived to listen to the evening's events, while Gamage's department store in Holborn invited its employees to stay behind after the doors had closed for the day to experience the great curiosity of the age for themselves.

Meanwhile, back in Chelmsford, the celebrity on whom the success of the exercise depended, had swept through the town like a barquentine in full sail and arrived at New Street in the manner of royalty.

At the dawn of the 1920s Dame Nellie Melba was about as famous as famous got. The Australian-born soprano may have been approaching her 60th birthday but she was still at the pinnacle of her vocal gifts. Made a dame for her work raising money to support Australian soldiers during the war, Melba had returned to Britain in 1919 and installed herself once again at the Royal Opera House in Covent Garden, to widespread acclaim. That night, however, instead of rattling the fillings of a London audience in black tie and ballgowns she would be singing into a glorified telephone handset in a draughty workshop surrounded by cables, valves and dials, and regarded earnestly by frowning men with their shirtsleeves furled to the elbows.

The idea to have opera's most recognisable voice sing over the airwaves came from Tom Clarke, an experienced newspaperman who'd risen to become assistant to Lord Northcliffe at the *Daily Mail*. Like many wireless enthusiasts Clarke's interest in the potential of the medium had been piqued as a signalman during the war. He had cultivated and maintained a friendship with Arthur Burrows, the head of publicity at Marconi, who was also convinced that the future of wireless lay beyond utilitarian uses like ship-to-shore messaging. Clarke persuaded Northcliffe that becoming associated with the new technology could be beneficial to his newspaper. Before the war Northcliffe had held huge influence. Indeed, the virulent anti-German propaganda published in his newspapers had helped to stoke the rush to war, and during the conflict Northcliffe was made Director of Propaganda by the Prime Minister David Lloyd George (a post in which he was so effective that a German battleship was dispatched specifically to shell his house at Broadstairs on the Kent coast). However, a subsequent falling out with Lloyd George had coincided with a postwar waning of Northcliffian influence. If, however, they could persuade the greatest soprano of the age to travel to the Marconi headquarters in Essex and broadcast a performance into the ether, one that could be received across Britain and beyond, then who knew what benefits there might be? It was also, Clarke knew, probably a good idea to be in on the development of the new technology from the start anyway.

At first Melba was not remotely interested, not even tempted by the whopping £1,000 fee Clarke dangled in front of her (at the time the average house in London cost roughly half that). Her voice, she felt, was not something with which to indulge the vulgar passions of a few hundred wireless enthusiasts and their 'magic playboxes' as she called them. This was, after all, a woman so famous throughout the world there was a dessert named after her.

Clarke and Burrows were persistent, however, and eventually the great dame was won over. She arrived at Chelmsford by train in the afternoon accompanied by her son, daughter-in-law and her two piano accompanists, Frank St Ledger and Herman Bemberg. A white Rolls Royce stood on the station forecourt to drive the prima donna the short distance to the Marconi offices where she dined on chicken, the unleavened bread she favoured and sipped champagne in the swanky surroundings of the Directors' Luncheon Room. After lunch she was escorted around the works by Burrows in order to give the great soprano an appreciation of the 'magic playbox' system and how it worked.

In the years prior to 1920 enthusiasm for listening to a wireless set far outstripped the entertainment value of the signals whizzing through the atmosphere. Marconi had some powerful transmitters set up at Chelmsford, two 450-foot masts that towered over the New Street works, and the company had conducted broadcast tests under the call sign MZX, but mostly these consisted of little more than a male voice reading out material not exactly designed to thrill the listener. Listings of London train termini and the routes they served, as well as highlights from the Chelmsford railway timetable, wouldn't have exactly been gripping material for people at Chelmsford station let alone a bright-eyed enthusiast pressing headphones to his ears in Ashby-de-la-Zouch to see what might be coming through the ether.

Broadcast radio seems so obvious to us now. If you had the capacity to send voices and music through the air then who wouldn't devise the breakfast show, the mid-morning phone-in, notable people selecting the records they'd take to a lonely desert island, congestion updates from the Hanger Lane gyratory system and, well, *You and Yours*? Yet when wireless telephony was being pioneered its uses were mainly concentrated on passing brief messages over long distances. These were specific, point-to-point messages of instruction, information and occasional requests for assistance that were entirely irrelevant to most of the people equipped to hear them. The notion of sending out broadcast material designed to be heard by an audience rather than an individual was slow to catch on. Audiences, the prevailing belief system laid down, gathered in one place, generally contained by four walls and a roof: the theatre, the music hall, the church, the concert hall, the opera house. Audiences converged on the performance.

Radio would turn that basic, second-nature human ritual completely on its head. Centuries, if not millennia of human interaction, were about to be entirely subverted: here was a technology that placed the performance entirely in the realm of the audience. From sermons to opera, knockabout comedy to drama, every aspect of audiencehood was about to swap poles. People wouldn't even have to leave their homes. Radio was about to cause arguably the greatest human cultural shift since the Ancient Greeks started building amphitheatres, and events in an Essex town on that warm day in the early summer of 1920 would be the biggest step forward yet. The potential was mindblowing. Even with the relatively small number of people able to tune in that night, it was still by some distance the largest audience for whom Nellie Melba had ever performed.

Having toured the offices, workshops and laboratories of the New Street works Burrows showed Melba and her entourage outside, where they all craned their necks to see the tops of the transmitting masts. Burrows, with an expansive sweep of the hand, explained to the soprano that her voice would be carried from the very tops of those masts out into the wide world beyond. There was a short pause as Melba looked up at the masts and then turned to look at Burrows.

'Young man,' she said, 'I am Dame Melba. If you think for one moment that I am going to climb up there I'm afraid you are very much mistaken.'

When the time came to prepare for the broadcast Melba found her surroundings to be a little more dignified than a ladder up a mast, but not much. The intention had been to set up a studio in a grand room in the main building, with cables run from the workshop to Melba's microphone. But during a test of the apparatus a power surge blew the modulating valves on the main console and sent a searingly hot charge along the cable which caused the microphone to explode. Marconi's engineers picked splinters of Bakelite from their hair, looked down beneath the smoking cable at the long, snaking scorch-mark across the carpet, and accepted that the broadcast would be made from the draughty experimental workshop.

Floors were swept, workbenches tidied and walls hurriedly whitewashed but, in terms of elaborate staging, the vast New Street experimental shed still fell a little short of the great opera houses of the world. Undeterred and ever the professional, Melba made the best of it. She toed the thick piece of carpet laid on the concrete floor to help deaden reverberations and announced, 'We'll be getting rid of this thing,' but, once the offending shagpile had been rolled up and spirited away she nodded tight-lipped approval and stepped outside to pose for photographs with the Marconi executives.

Meanwhile in Canterbury, Exeter, Banbury, Derby, Hull and everywhere else there was a receiver, headphones were donned, speaker horns angled, power cranked up, dials tuned to the 2,800-metre waveband assigned to the Marconi company by the Postmaster General, and an expectant hush fell, underpinned by the static hiss of the empty atmosphere.

At 7.10 p.m. a male voice called, 'Hallo, hallo, hallo!' and after a short pause the unmistakable sound of Melba's voice trilled up and down an arpeggio. There was a further pause – apparently a photographer's flashbulb going off had startled the chief engineer into thinking something had gone pop and he'd shut down the equipment – and then the same man's voice, possibly that of Arthur Burrows himself, announced, 'Hallo, hallo, hallo! Dame Nellie Melba, the prima donna, is going to sing for you. First in English, then in Italian, and then in French. The Marconi Company apologises but it has no control over atmospherics.'

A microphone – essentially a telephone handset with some minor alterations and a protective horn around it constructed from an old cigar box – was suspended from a rickety apparatus built from a hat stand. Melba stood straight, pushed her shoulders back, turned to Frank St Leger on the piano and nodded at him to begin.

She opened with 'Home, Sweet Home', then Puccini's 'Addio' from *La Bohème*, at which point St Leger vacated the piano in favour of Herman Bemberg, who accompanied Melba on his own composition 'Nymphes et Sylvaines'. As the last notes died away, the announcer cut in with another, 'Hallo, hallo! We hope you have enjoyed hearing Dame Melba sing. Good night!'

A technical hitch, however, had meant that most of the final song had been lost. Captain Henry Round, one of Marconi's most trusted engineers who had been with the company for nearly 20 years, fixed the problem and hurried from the transmitting room into the makeshift studio where Melba was about to step away from the microphone, her obligations fulfilled. Thinking quickly, Round addressed her. 'Madame Melba,' he said, with a respectful inclination of the head, then, clasping his hands together added, 'The world is calling for more.'

The soprano raised her eyebrows slightly at the prospect. 'Really?' she said. 'Shall I go on singing?'

'If you would be so kind,' said Round, backing away and bowing.

Nodding to Bemberg at the piano, Melba sang another of her accompanist's compositions, 'Chant Vénitien', repeated 'Nymphes et Sylvaines', and finished with a rendition of 'God Save the King'. The whole performance had taken a shade over half an hour.

With the return of the ambient hiss, listeners across Britain and beyond began to digest what they'd just heard. The Derby Wireless Club had had a few technical problems of their own at the start of the broadcast and missed the opening song but after that reported they missed not a note, trill or run. In Berwick Geoffrey Gore was awed by the experience, discussing animatedly with his fellow listeners how glorious the whole thing was, 'coming as it were from nowhere, out of space'. He noted that the sound had been clearer and more distinct than that of an ordinary telephone. At the Alexander Theatre in Hull Charles Greave couldn't suppress a cheer. So delighted had he been by the broadcast that he stayed at his equipment for a couple of hours afterwards, picking up Morse messages from Belgium and transmissions between ships. At one point human voices broke through, one a maritime wireless engineer asking the cost of sending a particular kind of message and receiving the reply, 'four and ninepence', an exchange almost as thrilling to Greave as Melba's recital.

On the bridge of the White Star liner *Olympic*, moored at Belfast harbour, there were handshakes among the captain Sir Bertram Hayes, his electrical manager named McKeown and a number of Harland and Wolff grandees who had gathered to hear the broadcast. Royal Navy ships at anchor off Gibraltar and Malta reported receiving the programme, and when the steamer *Victorian* docked at Liverpool from Quebec five days later its captain reported hearing the recital just over a thousand miles away in the Atlantic Ocean. Reports came in from Warsaw, Copenhagen, Stockholm, Madrid and Berlin that the recital had been heard loud and clear, while at the Eiffel Tower the reception was so good engineers were able to make a gramophone record of the performance. Across the city the French Radio Electric Company had set up a large amplification horn that had broadcast Melba's voice through a window into the street for the benefit of a large and enthusiastic crowd.

Yet for all these great distances the true magic of the Melba broadcast lay in those hunched over their crystal sets, headphones on, hearing the finest operatic voice of a generation coming through to their homes, their shops, or their sheds. People like Mr Bentley of Comberford Road, Tamworth, who reported that the prima donna's voice had been perfectly received, and the members of the Gloucester Wireless Society who overcame some less than favourable atmospheric conditions to enjoy the performance immensely – and Mr A. Priest of Somerset Road, Portsmouth, whose ordinary loose-coupled crystal receiver, the most primitive form of radio receiver and one easily assembled at home, afforded him almost perfect listening.

Mr Priest's crystal set would have comprised just four basic components: an antenna to pick up the radio waves and convert them into energy – a weak signal, but one strong enough to power the set without needing mains electricity; a piece of crystalline material that demodulated the radio wave to turn it into audio; a coil to effectively tune the device; and an ear-piece to enable the listener to hear the output. It was – and remains – the most basic form of radio but one that really captures the magic of the medium. Four basic components combining to decode the inaudible electro-magnetic vibrations carried through the air.

It's almost impossible today to imagine the thrill that must have overcome listeners to that Melba transmission. Most, if not all, will have come to wireless telephony through a scientific interest, but on that clear January evening science travelled way beyond the sending and harnessing of electrical impulses and radio waves and enmeshed itself with art. Until that night, to hear Nellie Melba sing live one required the opportunity and means to attend the world's most exclusive opera houses, yet here she was singing for anyone and everyone – people sitting in their homes, gathered in local electrical wholesalers and city centre department stores, on the bridge of an ocean liner.

Radio lent a new democracy to performance. There were no doors or turnstiles, no tickets, no private boxes, no 'sold out' signs and no dress code – not even a bus or train ticket was required. The greatest voice of the era was sent out to be heard not just by those able to attend the opera, but by anyone with the most basic equipment designed to draw the sound from the air. The number of people in Britain who owned receivers was still in the very low thousands at this stage, but they were people who could be found in high street shops and suburban terraces across the nation. Mass listening was still a good way off, but the Melba broadcast laid down solid foundations for the democracy of radio.

It also brought a message of peace. The men listening that night who'd served in the war would still have vivid memories of a conflict that had finished barely eighteen months earlier. Most will have spent months, even years, in trenches that never moved, their very existence governed by strict demarcations, boundaries that couldn't be crossed for the risk of certain death. Yet here was a science that respected no borders; one that sent words and music sailing and swooping invisibly through the air across land and sea, that couldn't be captured, altered or even slowed. Here, at last, among the fallout and traumatic memories of war, among the grief for those who hadn't come home, and among the simmering industrial discontent of those that did, was something positive, something magical even. Underpinning the excitement and exhilaration of that night there was a unifying feeling of hope. The process by which Nellie Melba selected her programme is lost for ever now, but beginning with 'Home, Sweet Home' gave the broadcast extra poignancy. 'A charm from the skies seems to hallow us there,' she sang of home. Here was a message of peace to carry forward, a new science, new possibilities and, above all, new hope.

Nellie Melba's recital wasn't the first broadcast in the world sent out for general consumption – a Canadian wireless enthusiast named Reginald Fessenden claimed to have broadcast a wax-cylinder recording of Handel, played the violin and sung a couple of songs into the ether from his home in Massachusetts on Christmas Eve 1906; and in the Netherlands Hanso Idzerda had a valid claim (see 'Radio Lessons in Hilversum', page 298) – but it was the first designed specifically to draw an audience. The image of one of the world's great operatic divas standing in a draughty workshop singing into a cannibalised cigar box hanging from a hat stand was a faintly ludicrous one, but it remains one of the most significant landmarks in British popular culture.

No recording survives, the gramophone record produced in Paris seems to have vanished, and that's for the best. Instead we are left to imagine the broadcast and its content from a few surviving contemporary accounts of people who were there and people who were listening. Either way, the greatest soprano voice in the world was carried across land and sea and headed for the stars, was lassoed by the coils and receivers of a band of pioneering enthusiasts, then faded to nothing, becoming subsumed among the eternal natural noises of the earth and the air.

9

The Bridle Path to Glory: the Birth of British Broadcasting

'Hallo, CQ, hallo, CQ. Two Emma Toc, Writtle testing, this is Two Emma Toc, Writtle testing. Hallo, CQ, hallo, Ash, hallo, Ash, are the signals OK? No they're not? Wave your hand if it's all OK. No waves? No waves at all? Hallo, CQ this is Two Emma Toc, Writtle testing.'

The voice had the kind of refined diction that would go on to characterise the BBC, but this was fast talking shot through with unbridled enthusiasm. The broadcast quality wasn't great, thin-sounding in its own right but rendered tinnier than a Cornish mine by the fact it was coming through my iPhone. Also, I was standing on a muddy bridle path in Essex trying to convince myself I was a doing A Significant Thing.

I'd left the car parked by Writtle village green, where some men with a cherry-picker were putting lights on a giant Christmas tree. It was one of those wintry days that make even the most pastoral surroundings a dull combination of greys, dark greens and browns, where the dampness seems to hiss in the air and the cold somehow starts in your bones and works its way outwards.

Walking along the edge of the green I veered to the right by the duck pond and came to a junction overlooked by a restaurant whose modern signage couldn't mask the fact that it clearly used to be a pub. Well, in fact I already knew it was once a pub, that it used to be called the Cock and Bell and that some of Britain's earliest radio programming had been devised in there. In fact, the pub's piano had reputedly been pushed along the exact journey I was making in order to accompany some of that programming.

Crossing the junction I set out along a road lined with some frankly gorgeous houses, all thick walls and windows latticed in little diamonds whose front gardens would be a blaze of floral colour come the summer, and turned right at a pair of pink cottages. I emerged onto a street of 1970s bungalows and houses of pale brick and white weatherboarding, followed the road round to the left towards a clutch of modern maisonettes and kept straight on until the road came to an end at a pair of bollards. Ahead of me was a bridle path of the kind you see right across Britain, a tree-lined track with fields either side made mud-shiny by hooves and footfall that led off arrow-straight into the distance. This was the place I was looking for. A short way along the path I stopped, looked up and down to see if there was anyone about, pulled out my phone, called up a very old recording and thumbed 'play'.

There are quite a few contenders for the birthplace of British radio, but the area around an Essex bridle path leading through a new housing estate a mile or so west of Chelmsford has a stronger claim than most. The thought of a group of young men pushing a piano from the Cock and Bell all the way here – an easy stroll for me but a little more challenging with the addition of a heavy cabinet of wires and ivory – is an example of all that was good about radio in Britain, certainly in its early days. It spoke of passion, enthusiasm and a sense that anything was possible, of being at the dawn of something enormous, so extraordinary you couldn't even imagine where it might lead even though you were on the outskirts of a tiny village in Essex putting your shoulder to an old pub piano with nicotine-yellow keys. Most of the young men manhandling the instrument would have been through at least the latter part of the war, and here, in the countryside, they must have gasped at the sense of freedom and excitement about the future a world away from the mud and lice of the trenches of Flanders. It was an image that shouted 'youth', the youth of the men involved and the youth of the medium of radio and all the possibilities contained in both.

It's also arguable that the BBC was born here: many of the men laughing and cursing as they tried to keep the piano on the straight and narrow would go on to be leading lights of the Corporation's early years as a direct result of their Writtle activities, laying the foundations for one of Britain's greatest ever global exports and achievements. And there I was, nearly a century later, using technology those lads could only have dreamed of, filling the Writtle air with the words of the man who was the driving force behind early British broadcasting.

This was the voice of Captain Peter Pendleton Eckersley, the Mexican-born son of a railway engineer and cousin of Aldous Huxley who would eventually become the BBC's first chief engineer. When the Marconi company sent some of its brightest young things out to the old hut at Writtle to experiment with new broadcasting technology, it was Eckersley who would become Britain's first regular presenter, broadcasting half an hour of often anarchic radio combining patter, gramophone records and frequently hilarious one-man sketches. In the recording I was playing to a couple of disinterested magpies and a horse the captain finished up by moving from the microphone to the piano and caterwauling an operatic aria, pretending it was a live link to a famous tenor in Rome.

Eckersley's story is a remarkable one. From a glorious and carefree 1920s he fell into a rather ignominious 1930s, and its roots were here, on the edge of a scenic village in rural Essex.

The Chelmsford experiments, of which the Nellie Melba recital was by far the highest profile, came to an end in November 1920 by order of the Postmaster General. There had been complaints, he said, that the broadcasts were interfering with military and emergency communications. The pilot of a Vickers Vimy, for example, flying from Calais to Lympne in thick fog had been unable to find vital directional and weather information over his radio because the Marconi station was playing gramophone records on the same frequency.

Britain's wireless amateurs, numbering several thousand by this stage, were put out. Opera superstars and the latest crooners made for much better listening than the few snatches of Morse code between ships that had hitherto been the only fare on which they could eavesdrop. Radio as a possible medium of entertainment had been a quirky by-product of its more utilitarian purposes, but there was clearly an audience for it. So much so that 63 radio societies held a conference in March 1921 to discuss the possibility of setting up a dedicated station for entertainment and information purposes that wouldn't interfere with the requirements of airborne coves in goggles picking their way through a pea-souper over the briny. At the end of the year the Marconi company presented a petition along with proposal to the Postmaster General, Frederick Kellaway, for permission to make regular broadcasts, one evening a week, for half an hour or so.

Kellaway was no Luddite: a former newspaper editor from Bristol who served for 12 years as Liberal MP for Bedford, he was in his first year as Postmaster General and so taken with wireless that when he gave up politics after losing in his seat in 1922 he joined the board of Marconi. A keen deer-stalker despite his wooden leg, Kellaway would be another credited as 'the father of British broadcasting' in his *Times* obituary. He agreed to the proposal and granted Marconi a broadcast licence but with strict conditions. The broadcasts would go out between 7 p.m. and 7.30, once a week, on any chosen day except Saturday or Sunday, and would be limited in power to one kilowatt. If the transmission was found to be interfering with official channels in any way the broadcast had to cease immediately: a member of staff should be at a receiving set at all times during the broadcast in case the Post Office sent a message to desist. There should be a break of three minutes after every ten minutes of broadcasting to allow any such message to be sent from London.

The company notified its wireless experimentation team out at Writtle, marshalled by Eckersley, that it would be responsible for carrying out the broadcasts, and from 14 February 1922 programmes went out on the station under the call sign 2MT – Two Emma Toc in the phonetic alphabet of the time. The programmes would run for nearly a year and set the tone and precedent for all the British broadcasting that would follow.

The Writtle hut was originally used for experimentation in ground-to-aircraft transmissions, and most of the 2MT staff had served in the wireless section of the Royal Flying Corps during the First World War. Conditions were spartan: the hut was on a floodplain downwind from a sewer, there was no mains electricity, no running water, a single coke-burning stove served to heat the 80-foot-long building, and toilet facilities consisted of an old speaker horn shoved into a bush outside.

Once the main Marconi broadcast equipment had been moved from Chelmsford to Poldhu and Derrygimlah, Writtle was the only site near the Chelmsford headquarters technically equipped for the 2MT broadcasts. Eckersley was in charge of a team of seven men and one woman, Lizzie Beeson, the daughter of the landlord at the Cock and Bell and a trained secretary. After a few brief test transmissions of Morse code the Writtle crew was ready to broadcast.

It almost didn't happen after a glass condenser exploded five minutes before transmission was due to begin, but some quick repairs meant that a few Morse signals were sent out ahead of a short programme of gramophone records featuring the singer Robert Howe, played on a wind-up HMV gramophone with the microphone held in front of the horn, and straightforward announcements of the artist's name, the title of the song and the record company that had produced the disc.

Eckersley was not directly involved in the early weeks, deciding he was more use going home to listen in order to provide informed feedback instead of standing over the gramophone with the rest of the crew – until one day in March 1922. The customary planning meeting in the Cock and Bell had gone on longer and been a little more, ah, convivial than it usually was, and when someone eventually realised the lateness of the hour the team had to hotfoot it down the lane to make it back to the hut in time to warm up the equipment.

Eckersley would later recall leaving the Cock and Bell that evening with 'a certain ebullience' that instilled 'a less formal attitude to the microphone than was customary'. He was hammered, in other words, and realising he wouldn't make it home in time to listen to the evening broadcast went back to the shed with the others and decided he'd have a crack at presenting the links between records that night himself. A new and vital exuberance was thus introduced to British radio; broadcasting would never be the same again and it was all down to a bunch of cheery drunks in a shed. Despite making his broadcasting debut well alight after a long afternoon in the pub Eckersley proved himself instantly to be a natural broadcaster. We all think we're confident, clear and funny after a few drinks but Eckersley actually was, even if he didn't quite get around to playing all the records sent from London for the purpose by Chappell of Bond Street, and even if he forgot entirely to stop for the required three minutes in every ten in case the GPO needed the frequency. He forgot to stop at all, in fact: the broadcast was due to finish at 8.30 p.m. but when the clock struck nine Eckersley was still reminding himself of funny stories and had, indeed, actually started to sing.

In the cold light of the following day a slightly fragile Writtle team gathered to discuss the previous evening's broadcast – a discussion punctuated regularly by Eckersley interjecting, 'Did I really say that? Really? I did? Good heavens . . .' – and awaited what they assumed would be a deluge of complaints and a comprehensive official bollocking. In fact, of the 50 postcards the team received that week only one was negative, from Marconi's head of publicity Arthur Burrows who, like Eckersley, would move on to the BBC and become one of the most significant pioneers of broadcasting himself as the corporation's first head of programming, but who that day pronounced himself shocked by the frivolity.

Eckersley was nevertheless a hit. His natural, irreverent style was perfectly suited to the new medium, one that had no rules, no traditions and no boundaries. With each passing week the (more sober) 2MT broadcasts began to sound more and more like the radio we know today: the Morse test signals were dispensed with, items of news were given out and, occasionally, singers would travel up from London to perform in the draughty shed at the end of the muddy lane. Nora Scott, a noted contralto of the time, performed three songs live in May, a broadcast that was received as far afield as Aberdeen.

In October the team scored another radio first when they broadcast a performance of the balcony scene from *Cyrano de Bergerac*, six of them around a table passing the microphone between them for each piece of dialogue after a solitary run through holding spoons in front of their mouths to replicate the microphone.

In the meantime the Postmaster General had granted the Marconi company a second broadcast licence for a station to be set up at Marconi House on the Strand in central London, call sign 2LO, transmitting twice a week on Tuesday and Thursday evenings under the guidance of Arthur Burrows. It was a little more formal than its Essex counterpart – 'careful pomposity' Eckersley called it – but, better resourced and better positioned in the heart of London (with two vast antennae on the roof of the building) it was enough to see off 2MT. While performers had to take a train to Chelmsford and be transported along the leafy lanes and muddy tracks of rural Essex to stand in a shed and then be bundled back to the station to make the last London train immediately afterwards, the Strand allowed for a little more decorum when it came to big-name artistes.

The last broadcast from Writtle was made on 17 January 1923 and such was the brouhaha surrounding the growth of 2LO and its mutation into the British Broadcasting Company the previous year, the Writtle closedown was barely noted. The station had come in with a bang and gone out with a whimper.

Other broadcast licences had been granted to manufacturers of wireless equipment around the country, and new stations sprang up. There was 2ZY at Metropolitan-Vickers in Manchester, and the 5IT at the Western Electric company in Birmingham. Frederick Kellaway, keen to avoid the chaotic airwaves prevalent in the United States, where unregulated broadcasting was leading to wild fluctuations in the quality of reception and the quality of programming, had set out legislation in May 1922 for a system of broadcasting that would benefit the public in general and not simply subject them to the whims of individuals with the means to broadcast.

Having called a meeting at the Institute of Electrical Engineers in London between the major manufacturers – Marconi, Metropolitan-Vickers, General Electric, the Radio Communication Company, Western Electric and Thomson-Houston – the formation of the British Broadcasting Company was announced on 25 May 1922, the new organisation granted £100,000 start-up capital and further income derived from issuing ten-shilling receivers' licences. The BBC's first offices were at Magnet House on Kingsway in central London, and the existing 2LO studios would serve as the BBC's London station. The following day 5IT in Birmingham and 2ZY in Manchester were brought officially into the fold and others would follow.

Before long Eckersley joined the new organisation as its chief engineer where he spent a happy and successful time bedding in the fledgling broadcaster, improving the quality of the equipment at 2LO and helping to set up regional stations around the country. He became a well-known and popular figure, addressing wireless societies and beating the drum for radio as a medium with extraordinary potential. Eckersley's reluctance to kowtow to authority and his penchant for openly criticising the BBC caused him to lock horns with the company hierarchy, but he was tolerated on the grounds that nobody in the country could touch him in terms of technical knowhow and experience.

As the decade closed, however, things really started going awry for Eckersley. In 1928 he'd begun an affair with Dorothy Clark, known as Dolly, the estranged wife of the conductor and BBC musical director Edward Clark. Dolly had left Clark before her relationship with Eckersley began and gone to work for the BBC as a secretary, but Eckersley was married. The couple seemed to have no qualms in conducting their relationship openly – indeed, Clark almost seemed to approve, helping to arrange a visit to Switzerland for them ostensibly on BBC business. Eckersley's wife Stella was completely unaware, however, until Muriel Reith, the wife of the corporation's Director General John Reith, broke the news to her.

Reith couldn't bear the possibility of scandal and even sought the advice of the Archbishop of Canterbury as to what action might be appropriate. Eckersley promised to end the relationship and return to his wife, but the rapprochement was temporary and when he and Dorothy resumed their affair Eckersley was fired. A terse BBC statement was put out stating that he'd resigned, notable in its lack of acknowledgement of Eckersley's role in the BBC's early years, but Reith had dismissed his chief engineer in person because, he said, Eckersley had 'strayed from the path of righteousness'. Stella sued for divorce and the following year Eckersley married Dolly, who had in turn finally been divorced by her husband.

Eckersley went on to work for HMV and advised for a number of wireless manufacturers, but one of the most influential and important figures at the birth of British radio would never work in broadcasting again.

As the 1930s progressed his political life became more significant than his working life. Dorothy, the daughter of a suffragette and member of the Independent Labour Party, followed her mother into the ILP, a party whom her husband embraced to such an extent he was invited in 1931 to become a parliamentary candidate. Instead he pitched in his political lot with Oswald Mosley and the New Party he'd formed after leaving the main Labour Party around the same time, an organisation that funnelled Mosley's transition from socialism to outright fascism. Dorothy, meanwhile, had become so attracted to communism she counted the Soviet ambassador to London among her circle of friends.

By 1935, however, the couple had both become card-carrying, drum-beating fascists, a political swing to the far right cemented that year by a holiday in Germany. There, Dorothy said, she and Eckersley had seen a nation where the government got things done instead of just talking about it. It was vintage 'at least the trains run on time' stuff. She joined The Link, a coven of particularly egregious whackjobs convinced the world was being run by a secret Jewish-Masonic conspiracy, who wanted to build tangible links between Britain and the fascists in Germany. So enthusiastic a

LAST TRAIN TO HILVERSUM

fascist did she become Dorothy also collected memberships of the Anglo-German Alliance, the Imperial Fascist League and, realising Mosley's British Union of Fascists wasn't virulently anti-Semitic enough for her, eventually threw in her lot with William Joyce and his gibbering National Socialist League. Indeed, she became a close friend and occasional financier of Joyce and also a confidante of arguably Britain's most famous fascist after Mosley, Unity Mitford. Peter Eckersley, meanwhile, while not such an enthusiastic joiner of organisations – he stuck to the British Union of Fascists – was just as enthusiastic when it came to espousing the far right cause. He set to work planning under Mosley's patronage a rival radio network to the BBC, one that would beam its signal across the Channel from various transmitters in continental Europe, and also advocated a cable radio service to compete with the BBC that was eventually nixed by the Postmaster General. One evening Eckersley was actually turfed out of a dinner party at the house of art historian Kenneth Clark for being an 'indoctrinated Nazi'. Yet at the same time there are reports that around 1937 Eckersley was recruited by MI6 to help counter the radio propaganda coming from Germany. Either way the couple spent their holidays in 1937 and 1938 at the Nuremburg rallies.

In 1939, as the clouds of war gathered over Europe, Dorothy chose to send her son James, from her first marriage, to a school in Germany, where she happened to be visiting him when war broke out. They decided to stay put and, at some point early the following year, now in different countries for however long the war would last, the Eckersleys separated. Before 1939 was out Dorothy had become an English language announcer on the Goebbels mouthpiece station *Reichs-Rundfunk-Gesellschaft*, where she was later joined by James, who became a newsreader. She also persuaded Joyce to remain in Germany when war broke out, leading to his infamous propaganda broadcasts as Lord Haw-Haw. After the liberation Dorothy and James were arrested as soon as they stepped off a plane at Croydon aerodrome. Dorothy was sentenced to a year in prison for the enthusiastic assistance she'd rendered to the enemy while James, pleading the impetuosity of youth, managed to escape a prison sentence.

Peter Eckersley lived quietly after the war, dying in 1963 at the age of 71. Fascism had always seemed an unlikely political ideology for a man almost genetically opposed to authority, and especially for a man who'd fizzed with such joy and enthusiasm for life and freedom in the year he'd spent in Writtle. Pitching in with the blackshirts, travelling to the Nuremberg rallies and being tossed out on the street by friends for being a Nazi bigot doesn't seem to tie in at all with the man who made thousands of people laugh, who could poke fun at authority on the air and rail against it off the air. The argument that he saw in fascism an ideology that shared his desire to get things done doesn't really seem to wash.

Standing on that bridle path adjacent to the site of the old Marconi works I listened to the recording I'd found, recreated by Eckersley a few years after the Writtle broadcasts but still betraying the sheer excitement and possibility of those days, the thrilling future that he could see opening out before him there at the dawn of radio. Maybe his embrace of fascism was born out of a disappointment with the bureaucracy and conservatism of the BBC and the frustrations he'd found in its rigid structures. Maybe his wife's ardent advocacy of fascist dogma just rubbed off on him. The voice that played through my phone that chilly afternoon sounded too energised by life, too full of sheer joy at the world and the possibilities it presented to belong to a man who would wilfully and enthusiastically embrace the same joyless, murderous philosophy as Hitler and Mosley.

I headed back onto the road, passing the weather-strafed information board that's the only indication that something remarkable ever happened in this unassuming locality. The little estate of maisonettes is grouped around a winding cul-de-sac called Melba Court, named for the opera star who made that historic broadcast, but a broadcast that happened a couple of miles away in Chelmsford. The 2MT hut itself was later removed to a local school where it served as a sports pavilion for a few years, and now a section of the hut forms part of the collection at Sandford Mill, a museum on the outskirts of

Chelmsford. The Writtle village crest on the sign by the green features a small radio mast but otherwise the momentous events that took place here are largely unacknowledged.

I'm not sure what I was expecting – a life-sized bronze statue of some lads in worsted suits pushing a piano up the street, or some kind of grotto – but then maybe such bells and whistles aren't really a radio thing. Radio is probably the most unpretentious medium of them all, its household names able to walk the streets unmolested, its carriage from presenter to listener entirely invisible. It's such a part of our lives it's almost become an extension of everything we do and everything we know.

Playing that recording of Peter Eckersley in a lane in Essex on a freezing cold weekday morning, I felt I'd gained an insight into what radio was like before it was ubiquitous, a time when it was the preserve of a few thousand enthusiasts smitten by the potential to send and receive voices and music invisibly through the air itself. What nature those words might take hadn't yet been prescribed, and Peter Eckersley's unfiltered *joie de vivre* exemplified the freshness and youth of the medium, the sound of which I'd just brought back to the site almost a century later, to the blank disinterest of a horse and a couple of magpies.

10

A Slumbering Old Warrior:
Looking for Lord Reith

There is a moment during Lord Reith's 1960 *Face to Face* interview that captures the image of the man perfectly. *Face to Face* was a series of hard-hitting BBC interviews conducted by John Freeman with an impressive range of luminaries from Dr Martin Luther King to Tony Hancock. Freeman, a former MP, and his interviewee sat closely opposite each other in a dark studio, the camera framing the face of the subject for almost the entire duration of the half-hour programme in an almost Beckettian study: every blemish, every twitch, every expression, every bead of sweat was in plain sight.

Reith was in his early seventies, and while his features may have softened a little with age the harsh studio lighting brought out all the granite Presbyterian cragginess in his brow and jaw. His hair, in wiry tufts each side of his shining cranium, had greyed but his eyebrows remained dark and bushy like a pair of gathering thunderstorms. The fearsome scar from the First World War sniper's bullet that had opened up the left side of his face nearly half a century earlier puckered menacingly in the shadows of his cheekbone.

Halfway through the programme the normally direct Freeman launched into a rambling question, one you could tell he wasn't keen on concluding, wondering whether when Reith looked back at his time at the BBC he was conscious of making any errors of judgement or mistakes. For the first time a long moment of silence fell across the scene. Reith looked levelly at Freeman, unblinking, unmoving, the studio temperature falling tangibly until the answer dropped with the weight of a stone tablet.

'No.'

There can be few figures who have dominated a nation's broadcasting like John Charles Walsham Reith. His is an influence that endures like no media executive's before or since. The BBC we know today is still hewn largely in his image and when you walk the corridors of Broadcasting House you can almost feel his presence even though he left the building in 1938.

Reith was a man who barely knew what broadcasting was when, in 1922, he applied for the post of general manager at the new British Broadcasting Company. Over the next 16 years he oversaw its transformation into a public service, fought tirelessly for its independence from governmental influence, and put in place a strict moral code: from announcers wearing evening dress to there being nothing 'frivolous' broadcast on the Sabbath, a policy that would unwittingly encourage the growth and development of commercial broadcasting, which he abhorred.

He even gave his name to a word that entered the *Oxford English Dictionary*: 'Reithian'. Its definition there speaks of the desire to harness broadcasting in a manner that informs, educates and entertains (a phrase often attributed to Reith that actually originated in the United States), essentially a commitment to high standards that for many people has come to define a strict, joyless moral didacticism. It's easy to think of John Reith as a monolith in two dimensions: a flinty-hearted killjoy with an icy glare constantly searching for signs of moral decrepitude, always judging, always proscribing, always condemning.

Much of that certainly applies, but there was far more to Reith. He was a man of depth and surprising nuance, who sought to overcome an emotionally barren childhood by developing whopping levels of ambition propped up by an enormous ego, setting standards for himself that condemned him inevitably to a life unfulfilled whatever he achieved. He was an awful father and an abominable husband, exercising complete control over the lives of those closest to him and expecting adherence to his personal code of moral conduct, one that made the Victorians look like patrons of the Hacienda nightclub circa 1989. Yet his own life fell short of those standards time and time again, often by a considerable distance. Somehow out of all his complexity and hypocrisy came the BBC and the wider broadcasting landscape we know today. For all his faults and flaws – and there were plenty – John Reith was the right man emerging at the right time.

A couple of miles south of Aviemore, down a single-lane road, a bumpy puddle-strewn track and behind a screen of birch trees is an old chapel. It must have been a forbidding house of worship in its day with its thick grey stone walls and spartan layout, but it stands open to the elements now, the stone floor worn and slippery with lichen. When I visited there was a thin covering of snow in the churchyard halfway to melting and I crunched around the graves for a while before homing in on the one I'd come to see. I was inexplicably nervous, putting off the inevitable like John Freeman floundering for as long as possible before actually asking Reith about possible weaknesses. I found a grave clad in an iron cage, which I learned later is the last resting place of Seath Mor, a chief of the Shaw clan from the turn of the fifteenth century. The iron cage is to protect the five round stones that sit on the grave, with local legend dictating that if the stones are tampered with their goblin protector appears to wreak a terrible punishment on the perpetrators. When local youths began messing with the stones as part of a rite of passage the council clad the ancient warrior in an iron trellis. According to legend people walking in the forest of Rothiemurchus have come across a giant, terrifying figure emerging from the shadows issuing them with challenges. They say it's Seáth Mor but there's possibly another contender.

I left one slumbering old warrior for another. Having beaten the bounds of the churchyard I walked almost hesitantly towards Lord Reith's final resting place. A few weeks earlier, walking down George Street in Edinburgh, I'd gone on a whim to see the house where he'd spent his final months – a tall, narrow building at the end of a quiet terrace in which the elderly Reith had been granted a grace and favour apartment. It was a blustery day, all low scudding clouds and rustling leaves, and when I found the address the wind caught hold of a blue plastic recycling tub further up the street and sent it skittering along the cobbles until it came to a stop right at my feet. For a fleeting moment I thought about some of the programmes I'd made for BBC radio. It was, of course, pure coincidence that as soon as I arrived at Lord Reith's doorstep some invisible hand had seen fit to propel a receptacle for rubbish a good 30 yards along the street until it came to a halt right under my nose.

Reith's grave doesn't have quite the same level of mythology or ironwork as Seàth Mor's. It's marked by a small simple stone cross, no more than two feet high, the name 'Reith' carved into the plinth, the dates of the man and his wife and two words of Latin, *pulvis ipsorum*, meaning 'their dust'. No titles or honours, no fuss, no flourishes beyond two words of Latin that speak of humility. Short blades of grass emerged warily from the ledges of the stone and a carpet of fallen leaves was spread all around.

The chapel at Rothiemurchus is a beautifully quiet spot. The only sound was the hiss of the wind in the skeletal trees and the occasional rapid knocking of a woodpecker somewhere in the vicinity. For a man who holds such an important place in the history of modern culture – indeed, is one of the most important figures of the British twentieth century – Rothiemurchus is about as understated a place for eternal rest as you could imagine. Reith's ashes lie buried at the base of one of the walls of a derelict chapel in a churchyard so well hidden that you're not sure you've even found it until you have a hand on the gate. The previous day I'd gone into the Aviemore tourist office and asked where I might find it, and after much staff discussion was advised to return the following morning and speak to Seumas, who would probably able to help. Not only was Seumas, a kindly man with sparkling eyes between fans of lines that indicated a life of laughter, able to help pinpoint the old chapel he was able to tell me exactly where in the churchyard I'd find what I was looking for, having wheedled out of me the sheepish admission that I was there to see the last resting place of a long dead broadcast executive to whom I had no personal ties whatsoever.

It seemed an odd place for Reith to choose. On the face of it he had no more connection to the place than I did: he was born in Stonehaven, on the east coast of Scotland just south of Aberdeen, grew up and studied in Glasgow, spent most of his life in London and died in Edinburgh. So why choose to have his ashes buried in an abandoned churchyard at the heart of the Cairngorms?

The answer lies in his childhood. Reith was the youngest of seven children – and the youngest by ten years at that – of Rev Dr George Reith, a prominent Free Church of Scotland minister. Like his son, Reith senior demanded high standards from all those around him and demonstrated little in the way of paternal affection.

For a few weeks every summer the Reith clan would swap their Glasgow manse for the minister at Rothiemurchus – ostensibly a holiday, but such feckless self-indulgence as holidays didn't sit well with the Reverend Doctor: he would become the minister of the parish for the duration of their stay and continue with his voluminous output of religious writings just as he did in Glasgow.

The age gap with his siblings meant that Reith spent a lonely childhood, but in the Cairngorms he found blessings in solitude. As soon as the family had disembarked from the train at Aviemore he would run to the end of the platform for a first proper panoramic glimpse of his beloved mountains not restricted by the frame of a railway carriage window. He could name them all by sight too. He'd spend the weeks at Rothiemurchus roaming the countryside and, as he got older, climbing the mountains. And it was at Rothiemurchus that Reith experienced the moment of self-revelation that transformed his life. When he was 17 he climbed to the top of Ben Macdui, the second-highest mountain in these islands after Ben Nevis, and, as he stood at the top surveying the landscape, was suddenly infused with and engulfed by a certainty that he was destined to achieve great things. It was a religious moment for Reith, looking out across the wonders of God's creation and feeling as if his maker was instilling in him a great purpose to be achieved in his name. He didn't know what it was but he knew that, whatever happened, the youthful John Reith would not grow to be a man of mediocrity.

All the Reith children suffered emotional neglect from their parents, whose prime focus was God rather than family, and John, so far behind in age, was left more isolated than the rest. He craved affection, especially from his mother whose company he was granted for just an hour a week, and craved at least the respect and attention of his father. It never truly came and, although as he grew older the pedestal on which he placed his father grew ever larger, it took those long summers striding alone through the heather and the hills for Reith to discover the person he was or, more accurately, the person he'd decided he was going to be. Where the great drive that fired his life and powered that gargantuan ego came from is all but impossible to define exactly, but the quest to win his father's approval and the majesty of the highland landscape were certainly important ingredients in the mixture.

His father's refusal to permit him to continue to higher education was also a factor in Reith's later ambitions. After being expelled from his academy school in Glasgow for bullying, Reith had been sent to a boarding school in Norfolk where, again, he was destined to be the perennial outsider – the only Scot and the only Presbyterian; his height also making him stand apart (he was over six feet tall by the age of 15). While Reith had nurtured thoughts of Oxbridge – despite his less than stellar school career – his father chose for him instead an engineering apprenticeship at a Glasgow railway company. While his intention was that his youngest child should have a trade, George Reith had inadvertently thrown down a challenge to his son to prove him wrong and even impress him. It was shortly after this decision was handed down to him that Reith experienced his mountaintop epiphany.

In 1914, at the age of 24, Reith moved to London. In his published writings he would say that London was the inevitable destination for any man of serious ambition – which is true, but the main draw of the metropolis was a 17-year-old-boy named Charles Bowser. The Bowsers had been near neighbours of the Reiths in Glasgow and the engineering apprentice had first got to know Charlie in 1912. Reith recorded in his diary that he found Bowser to be handsome with 'awfully pretty eyes' and developed a full-on infatuation with the boy. They took long holidays in the highlands during which they shared beds and baths and carved their initials into trees together, and Reith was left bereft when the Bowsers, who had strong misgivings about their neighbour's influence over their son, moved to London. The two would see each other as often as they could until Reith packed his bags and set off south for the metropolis himself.

Having served as a territorial soldier with the Lanarkshire Rifles Reith was one of the first in line when the war started in the summer of 1914. He took two photographs with him, one of his father and one of Charlie Bowser. In October the following year, while inspecting overnight damage to some trench defences, he was hit in the face by a sniper's bullet. At six foot six Reith was literally head and shoulders above most of the men and often displayed a foolhardiness in the trenches that smacked of a man convinced his destiny lay well beyond the war: he was well known, when faced with walking along a trench filled with soldiers, to simply hop up onto the parapet until he'd passed the obstruction. On one occasion, rather than go in search of a latrine, he climbed out of a trench and walked 20 yards into no man's land to relieve himself.

As he lay wounded awaiting medical attention he wrote two brief notes, one to his mother assuring her he was all right, the other to Charlie, advising him to keep his chin up. After a long spell in hospital Reith spent two years in the United States as an inspector of arms contracts before returning to take up a post overseeing the construction of submarine barrages off the Sussex coast. He found Charlie a job nearby, and one afternoon the pair set off house hunting in a regimental car driven by a young volunteer named Muriel Odhams, a woman whose fiancé had been killed earlier in the war. This chance encounter would soon develop into a curious *ménage à trois*. Reith and Bowser shared a house and Muriel would spend nearly all her free time with the two men, taking long walks over the Downs arm in arm with both of them, much to the consternation of her mother. For Muriel it seemed inevitable that she would marry one of the two men – preferably Charlie, if she was pressed on the subject – and eventually on one of these walks the pair instructed her to go on ahead while they fell into a deep discussion. After a few minutes Reith bounded forward and placed a ring on her finger. It was a ring he'd shown her once before, explaining that while most of his love was reserved for his parents, he'd made room in his heart for Charlie and, who knows, maybe one day he'd love her too. Then he'd put the ring away again. Even for the social mores of the time this was a monstrous display of arrogance, but on the day Reith finally bestowed the ring and his plighted troth on her she still agreed to marry him (Reith's father was typically blunt about his son's engagement, asking if Muriel came from 'healthy stock' and was 'thoroughly healthy herself', the soppy old romantic).

The barrage work in Sussex ended and John went north to take up a job at an engineering firm in Coatbridge, naturally taking Charlie with him and renting a house in which they shared a bedroom. Muriel, meanwhile, remained in Sussex until they married in Brighton in 1921 and she was transplanted to a new home in Dunblane, where the best bedroom in the house would be kept permanently made up on the off chance Reith's mother might visit.

Soon afterwards Reith decided that if their relationship was to continue Bowser needed a wife too and set about finding him one. That way, he surmised, Bowser could have all the conventional trappings Reith boasted while continuing their own close relationship unhindered. Selecting for Bowser a well-connected local girl named Maisie Henderson, Reith found himself put out when the couple began to show a genuine affection for each other. In 1922 Bowser, now in his mid-twenties and apparently no longer prepared to let Reith organise his life, informed his friend that once he and Maisie were married their relationship could not continue as before. Reith would never lose the bitterness of what he saw as the betrayal of the only person who had ever understood his complex and demanding personality. Even in later life Reith would become sullen and bad-tempered on Bowser's birthday, and he burnt more than 400 of his letters. While nobody knows if or to what extent their relationship had been a physical one behind the closed door of their shared bedroom, Reith's relationship with Bowser would always be the one that dominated his emotional life.

In the autumn of 1922, possibly in response to the end of his friendship with Bowser, Reith escaped to London again, taking up a position of secretary to a group of London MPs as they prepared to contest that year's general election. On 14 October, leafing through the *Morning Post*, an advertisement for a general manager at the newly founded British Broadcasting Company caught his eye, and from that moment the twin destinies of the man and the broadcast institution became inextricably intertwined. Reith passed through the interview without a hitch – an extraordinary feat of bravado and self-belief when you consider that he had not an inkling of what broadcasting actually was – and began work on 30 December 1922.

It was the perfect post for Reith. The BBC was a fledgling organisation in an entirely new field ripe to be shaped by a strong personality. Reith recognised in the company the potential to become an enormous influence on society, with access to the most influential and important people in the land. For a man with a staggeringly high opinion of himself, who felt that he was destined to shape great events and achieve great things, his tiny office at Savoy Hill – one so small he said he had to hang his hat on his walking stick in the corner because there was no room for a hat stand – was the perfect place to begin. No matter that he had no clue as to the nature of broadcasting, this was the job for him and he was the man for the job. He revelled in the control he had and the influence he immediately began to wield. Early in his appointment, having contacted the Archbishop of Canterbury to discuss the direction of future religious broadcasting and been told that the Archbishop had no means of listening to the BBC, Reith seized the opportunity to invite the Primate to dinner at the flat he and Muriel had taken in Westminster. On their arrival Reith switched on the wireless set and orchestral music filled the room. Impressed, the Archbishop asked Reith if the BBC ever broadcast piano music, of which he was a particular fan. Reith immediately telephoned the director of music and within minutes the dining group was listening to a piano rendition of Schubert's 'March Militaire'. When he saw the mixture of delight and awe on the face of the Archbishop of Canterbury Reith truly realised the influence his new post promised. Not a year earlier he was managing an engineering works in North Lanarkshire, now he was not only entertaining the Archbishop of Canterbury to dinner at his home, he was inspiring awe in the most powerful religious figure in the country.

In contrast to most self-aggrandising blowhards it was clear that Reith's arrogance was backed up by talent and ability. When newspapers, fearing the competition radio would provide in news coverage, insisted the BBC should pay to have its schedules listed in their pages, Reith responded by launching the BBC's own *Radio Times*, a publication still going strong nearly a century later. He overcame early problems with licensing to practically double the BBC's income at a stroke and, in 1925, wrote *Broadcast Over Britain*, a book detailing the early days of the BBC which set out much of the philosophy that would dominate the organisation in his image: editorial independence, not broadcasting entertainment for entertainment's sake, and promoting the key role of Christianity in society while lamenting the secularisation of the Sabbath. He oversaw the move of the company's offices from the cramped Magnet House and the old 2LO premises on the Strand to the slightly more spacious Savoy Hill, between the Strand and the Thames, and set in motion the process by which the BBC would become a public corporation at the beginning of 1927.

Perhaps Reith's finest hour came during the nine-day General Strike of 1926. While most newspapers ceased production for the duration of the industrial action the BBC kept broadcasting and became by far the most reliable source of news. Such was the power of the newspapers prior to the strike that the BBC had been prohibited from broadcasting news bulletins before 7 p.m., and even then could only use material sent to them from news agencies, so as not to affect the evening papers' circulations. That ban was lifted during the strike, and the BBC's news bulletins at Reith's behest remained entirely impartial, reporting the activities of the unions and the TUC as neutrally as they did the actions and statements of the government, much to the chagrin of those in the corridors of national power. Winston Churchill, then Chancellor of the Exchequer, lobbied the Prime Minister Stanley Baldwin in favour of the government taking over the BBC for the duration of the emergency, something Reith was able to avoid by pointing out to Baldwin that doing so would destroy any trust the public would ever have in the editorial independence of the BBC.

When Baldwin addressed the nation early in the strike he did so from Reith's apartment, with some of his statement even rewritten by the BBC's general manager while the Prime Minister was on the air. Reith even took over the reading of some of the news bulletins on the grounds that his regular announcers sounded 'too anxious'. This hands-on approach was fired by his determination to preserve the independence of his organisation in order that it might retain the trust of the public, without which the BBC would be worthless. It was something he would continue whenever the nation experienced the most dramatic moments of his tenure: Reith introduced 'His Royal Highness Prince Edward' when the erstwhile King Edward VIII addressed the nation on the subject of his abdication, and insisted on reading out the results of the 1929 General Election himself. When listeners called to complain he was going too fast, the head of presentation was forced to tap his boss on the shoulder as his giant frame hunched over the microphone and inform him of the listening public's criticism of his presentation.

'I will not read any slower,' Reith barked, 'and I will go on announcing.'

The head of presentation was fired the next day.

There were no such criticisms regarding the General Strike coverage, however, at the conclusion of which Reith read out a specially composed speech that ended with him reciting the words of 'Jerusalem' as the BBC orchestra swept in to accompany him. The BBC was inundated with thousands of letters complimenting its even-handed coverage of one of the most tumultuous moments of the British twentieth century. Those nine days can be pinpointed as the period that established the BBC as the bastion of independence it would remain for decades to come, and that was entirely down to Reith's implacable and inflexible style of management. He was knighted at the end of the year, a few days before the BBC became a publicly owned corporation under a board of governors rather than answering to shareholders, but his teetering ego prevented him from feeling honoured by the bestowment. An 'ordinary' knighthood, he opined, was 'almost an insult' considering what he felt he had achieved.

While Reith had been the instigator of the BBC's pioneering brand of public ownership – a hell of an achievement given he had to persuade the company's directors and shareholders to relinquish their roles and stakes in what looked to be a pretty hot financial prospect – he found answering to the Board of Governors far more exacting than the directors for whom he'd worked when the BBC was still a company. He'd been left almost entirely to his own devices when he started the job (he was just required to update the board once a month), but the BBC governors wanted weekly meetings, and there were frequent clashes between them and their authoritarian Director-General. So much so that in 1930 Reith decided he would be much better suited to the chairmanship of the Board of Governors, and when the Prime Minister Ramsay MacDonald appointed John Henry Whitley, a former Speaker of the House of Commons, Reith spent almost the entire funeral service for the Archbishop of Canterbury (which they both attended a few days later) glaring at him.

In 1929 Reith dealt with the Eckersley affair with perhaps surprising sensitivity, soliciting advice before giving his chief engineer the opportunity to finish his extramarital activities and return to his wife. It was only when the romance erupted again that Reith was forced to inform him he had 'strayed from the paths of righteousness' before adding, 'you are dismissed'.

In 1932 Reith oversaw the move to the specially constructed Broadcasting House in Portland Place – with the legend 'RECTORE JOHANNI REITH PRIMI' inscribed over the reception area recognising the corporation's first Director-General, or 'rector' as the closest Latin word would have it – and the same year was the first voice to be heard on the BBC's new Empire Service, broadcasting to the British colonies around the world.

Successful though Reith undoubtedly was, his management style and domineering personality didn't go down well with everyone. In 1930 one of the BBC governors expressed the view that Reith wouldn't be in the job much longer because he was 'quite mad', while another vocal critic from the boardroom asked him outright from where he had developed his Mussolini-like traits. The BBC continued to go from strength to strength, however, becoming a global bastion of honesty, integrity and truth. A fledgling television service began, something on which Reith was not keen at all (one contemporary noted that Reith was on holiday for all the early landmark moments at the birth of British television), and while Reith maintained his authoritarian manner at the BBC through the 1930s – and indeed expressed admiration for Hitler as a man who could get things done – he was beginning to grow restless. His burning sense of ambition had been doused to an extent by his leadership of the BBC but he soon began to feel as if his talents and energies might be better served elsewhere. In typical Reith fashion he felt that he'd done such a good job and implemented structures that were working so perfectly he'd left himself without enough to do.

The full reasoning behind Reith's departure from the BBC has never been made completely clear. The Prime Minister Neville Chamberlain let it be known that he wanted Reith to take over the chairmanship of the struggling national airline Imperial Airways. The BBC governors hinted that if he wanted to go they wouldn't stand in his way, and Reith himself had, for the previous three years, made ill-concealed noises about seeking fresh challenges. Newspapers had been speculating openly on who might succeed him (with the odd tongue-in-cheek suggestion that only God himself could possibly fill the Reithian shoes). When the end came it seemed to occupy a grey area somewhere between leaving on his own terms and dismissal. Reith himself struggled to explain the nature of his departure in later years, but on 19 July 1938 he walked out of Broadcasting House for the last time, with tears in his eyes, then drove to Droitwich where, at midnight and the conclusion of programming for the day, he ceremonially shut down the mighty transmitter himself. As he left he signed the visitors' book, 'J.C.W. Reith, late BBC.'

His departure and the ensuing years brought him no satisfaction. When war was declared Reith became Minister for Information in Chamberlain's government, and when his old adversary Churchill became Prime Minister he was moved to the Ministry of Transport and created a Baron to earn a place in the House of Lords before he was dismissed from the cabinet in 1942 in favour of a Churchillian Tory. When subsequently offered the post of Lord High Commissioner at the Church of Scotland, a job he would otherwise have coveted as one of which his father would have approved, Reith turned it down on the grounds that Churchill was a 'bloody shit'.

He wrote to an old friend at the Admiralty offering his services, his only stipulation being that he be 'kept busy' – by busy he meant taking on 'three times as much work as you can imagine anyone doing'. He ended up playing a key role preparing ships for the D-Day landings, becoming so intimately involved in the planning he was one of the few given top secret advance notice of the invasion.

After the war he assumed directorships of a number of telecommunications companies but still yearned for a job of real influence and standing that would stretch what he saw as his immense range of talents and capabilities. One job he coveted in particular was Viceroy of India, but when the post was given to Lord Mountbatten the new viceroy went inevitably into Reith's black book of grudges. And if anyone could hold a grudge it was John Reith (Mountbatten, for example, would always be a 'playboy' and a 'fraud').

He kept an eye on developments in broadcasting and rarely liked what he saw. When BBC television began showing greyhound racing in the early 1950s he described it as the most significant display of public depravity he had ever seen and when, in 1954, the government gave the green light to commercial television Reith compared its effects on society to those of the bubonic plague.

He grew ever more irascible and arrogant with the passing years, openly declaring to all who would listen that he should have been Prime Minister. And his attitude to his daughter and, in particular, her marriage was especially monstrous. When Marista Leith announced her intention to marry Murray Leishman, an amiable Church of Scotland minister, her father immediately placed his future son-in-law on his hate list. That list chopped and changed over the years but Murray Leishman was a permanent fixture, not because Reith disliked him personally but because Marista looked to be escaping his control. To introduce Reith to his future son-in-law the couple practically ambushed him at Euston Station as he headed for his car. As Leishman stood with his hand out, Reith ignored them both and made to fold his huge frame into the back seat until his chauffeur indicated the couple's presence. He eventually shook the outstretched hand, but maintained a stony, glaring silence. There was serious doubt as to whether he would attend the wedding: as it turned out he did so, but failed to acknowledge either bride or groom throughout the ceremony. It might be no coincidence that Murray Leishman eventually left the church and became a psychoanalyst.

As I left the churchyard and made my way back towards the puddled track I thought back to another moment from Reith's *Face to Face* interview. John Freeman had alluded to Churchill's description of Reith as being a man who was extremely difficult to get along with. Did Reith feel that was an accurate summation of his personality?

This time there was no silence and no hesitation.

'Yes!' he barked. 'Yes!'

11

Women Against the Tide: the Ballad of
Sheila Borrett

The chances are you've never heard of Sheila Borrett. Not many people have. Her BBC radio career was brief, a very long time ago and not exactly a proud episode in the history of the Corporation. She wasn't even called Borrett for very long. Yet Sheila Borrett was a trailblazer: one of the most significant figures in British radio history, whose contribution to that history comprises just one part of an extraordinary radio life.

When, at 6 p.m. on 28 July 1933, the nation's wireless listeners heard, 'This is London calling. Tea-time music today comes from the Hotel Metropole and it will be played by the Hotel Metropole Orchestra,' it should have been a standard piece of BBC continuity like any other. Instead that announcement and others that linked programmes for the rest of the evening were rendered under more scrutiny than any piece of British broadcasting in the short history of the wireless.

The voice belonged to Sheila Borrett, who had just become the first woman announcer ever to appear on British radio. With that simple flagging up of forthcoming music she laid the foundations for all the women broadcasters who followed – from Daphne Oxenford to Annie Nightingale to Sue MacGregor to Jo Whiley to Fi Glover to Lauren Laverne . . . they all followed Sheila Borrett.

Naturally, in the wake of her debut there was one particular section of the population who saw fit to immediately begin stroking their chins and giving their opinions: men.

'Her voice, described as "mezzo", struck me as deeply unusual,' said one critic, 'deep and of uniform timbre. There is a sombre quality about it.'

'Her voice is cool, businesslike and commanding,' said another. 'Her tone is authoritative, crisp and clear.'

Well, thank goodness for that. A little over three months after that first announcement, however, despite general approval of her work, Sheila Borrett was effectively hounded from the airwaves. It would be years before another woman's voice would be heard on British radio outside of dramatic productions or singing in variety shows. Sheila Borrett, however, didn't really give a toss. She knew that social conventions and the whims of men were no basis to live a fulfilled life – it was a lesson she'd learned in London to add to the one she'd learned halfway round the world. Sheila Borrett had no intention of allowing men to dictate the path of her life, not least because she was absolutely magnificent.

She was born Margaret Sheila Graham in the summer of 1905 at Harrow School, where her father Edward was a housemaster and a teacher of Latin and Greek. She was the youngest of three sisters, while her brother Francis was killed just a few months before the end of the First World War. Her mother Margaret came from noble Scottish stock, the Stewarts, and in interviews late in life, when she lived in Florida, Sheila would claim descent through her maternal line from Mary, Queen of Scots and King Kenneth I. Now, that might have been the case, or she may just have been cranking up the posh British schtick for her American audience, but whatever the truth was it showed how she intended to remain in control of her own narrative. She clearly identified strongly with her Scottish roots, however, always giving her nationality as Scottish despite never actually having lived in Scotland; and in her early acting years (as well as the second half of her life spent in the USA) she went by the professional name of Sheila Stewart. Whatever the truth behind her heritage the family was notable enough for Sheila and her sisters to be presented at court as debutantes in their teens.

The loss of her brother aside, her childhood seems a perfectly happy one spent mainly at the family home in Dorset where, with her sisters, she would act in plays and perform songs and sketches at local events. She would also claim in later life that she gave up studying at Oxford in order to pursue an acting career, but it seems it was marriage rather than the sock and buskin that diverted her from the dreaming spires.

Sheila Graham was 19 in July 1925 when her engagement was announced in both *The Times* and *Tatler* to Captain Philip Whitefoord, MBE, MC. Whitefoord was 11 years her senior and an army officer stationed at Fort William, Bengal, where Sheila's mother Mary had been born when her father, a distinguished military man, was stationed there. Whether Mary had been behind the engagement isn't certain but seems likely: the chances of romance blossoming spontaneously between a 19-year-old aspiring actress in England and a 31-year-old war veteran living in India are probably pretty slim without a firm maternal hand in the small of the back. Sheila and her mother set sail from Liverpool on 7 November 1925 on the *City of Exeter*, with Sheila, according the ship's manifest, intending to make her permanent home in India. Sheila Graham became Mrs Sheila Whitefoord on 21 December that year at the Church of St Peter, Fort William, Bengal, with her mother as her witness and no other family present. Mary sailed for England after Christmas, leaving her daughter in a strange country with a husband it seems she barely knew.

Sheila remained in India for less than three years. After spells living in Jhansi and Poona she arrived back in England in October 1928 – alone. Seven months later she filed for divorce from Philip Whitefoord, citing his adultery with a woman at the Craven Hotel near Charing Cross station, adding that it wasn't the first time he'd been unfaithful. Whitefoord didn't contest the suit and remarried within weeks of the divorce being made final.

Whatever happened in India the end of her marriage seemed to galvanise Sheila. She'd given the patriarchy a chance and it had been found wanting. It wouldn't happen again. She threw herself into the acting career she'd coveted since those early sketches and revues with her sisters and joined the company at the Old Vic for a while, as well as appearing in radio plays for the BBC. She also became involved in the fashion business, at one point around the turn of the decade touring Europe for a London fashion house in the company of ten models and their associated trunks and luggage on a trip that covered the continent from Rome to Scandinavia.

Still only 24 years old, Sheila was demonstrating an iron will and a determination to win back the wasted years she'd spent in India. In 1929 she even claimed to have raced cars at Monte Carlo wearing black from head to foot and driving a black car, finishing fifth. Like her royal Scottish heritage this story was told to an interviewer in the US towards the end of her life, so may need taking with a hint of reasonable doubt. According to Sheila, when she told her mechanic she wanted everything inside and outside the car to be jet black he pointed out that the driver's mirror would ruin the effect. 'Take it out,' she said. 'No one will pass me anyway.'

I couldn't find any reference to any women drivers at Le Mans or Monte Carlo before 1930, but I would put nothing past Sheila Borrett. The more I learned about Sheila the more I liked her and realised she was capable of pretty much anything she put her mind to.

In the summer of 1930 she married again, to Commander Giles Borrett of the Royal Navy, in a simple ceremony at the Savoy chapel close to the old BBC headquarters by the Thames. The Savoy was a rare venue where couples who had been married before could still have a church service; indeed, in *Brideshead Revisited* Celia tells Charles of her own wedding there that it was 'the place where divorced couples got married in those days', despite being 'a poky little place'. What the Borretts thought about the union between their naval officer son and the divorced actress can't be ascertained for certain from this distance, but we might permit ourselves a guess when we consider that Sheila's new mother-in-law attended the service dressed from head to foot in black with a black bulldog at her side.

The marriage seemed a happy one, happy enough that in April 1932 Sheila gave birth to a son, Jeremy. The following year she heard about a vacant announcer's position while performing in a play at Broadcasting House and decided to apply. 'I think they only gave me the job because I had fluent French and German,' she said later, of her new position, one that would pay her £500 a year (equivalent to around £25,000 today).

The intention had been to keep her identity a secret but once the news leaked out that the BBC had appointed a woman announcer it wasn't long until the press had identified 'Mrs Giles Borrett'. Journalists turned up on her doorstep and when 28 July arrived there were so many photographers, reporters and curious onlookers at the door to Broadcasting House she was forced to enter the building by a rear entrance.

Reactions to Borrett's announcing debut were overwhelmingly positive, if wincingly patronising. The *Daily Express* ran a spoof interview with Sheila's 18-month-old son 'on first hearing Mummy broadcast', while according to an unnamed 'authority on broadcast English' she had a 'pleasant and cultured voice'. Another apparent expert commented that, 'There were names contained in the announcements that might have made even a hardened male announcer pause, including Dvořák, Amanda, "Serenade Galante" and *paso doble*.'

Such astounding verbal dexterity soon led the BBC to permit Sheila to ascend to the blue riband of announcing: reading the six o'clock news. She did this for the first time on the evening of 21 August when, amazingly, her lady mouth didn't prevent her from coming over 'loud and clear' according to reports afterwards. It was also one in the eye for another alleged broadcasting expert who had commented after her original appointment that if Sheila was let loose on the news there would inevitably be items she would be simply incapable of reading well. 'You can't expect a woman to work up any enthusiasm about fat stock prices or a revolution in Timbuktu,' he'd said.

Her evening shifts, of course, had to be conducted wearing evening dress, in line with the bizarre BBC regulations of the time. If you happened to be on duty for both the afternoon and evening shifts you were required to change into evening wear during the six o'clock news unless you were actually reading it, in which case you had to change immediately afterwards. 'I was told very firmly that it must be discreet,' Sheila recalled later. 'I don't know whether they pictured me coming in in gold, naked to the navel, or what, but they were very firm that it must be very discreet.'

While her early announcements had been well-received – inasmuch as critics were impressed that her lady brain allowed her to get through a whole five-minute news bulletin without being distracted by thinking about pretty dresses – once Sheila began reading the news the tide began inexplicably to turn. There had been a smattering of complaint letters ever since her appointment, but as summer became autumn their numbers began to increase. When, on 14 October, she informed the nation that Germany had left the League of Nations, one male critic was convinced the task of relaying news of such import was a man's job, and a man's job only. 'The full gravity of the situation was not brought home by a woman's voice,' he harrumphed, having apparently been hoodwinked by Sheila's lady timbre into thinking he'd heard some tittle-tattle about recipes instead of important developments in international relations.

Shortly afterwards the BBC announced that Mrs Borrett was being relieved of her announcing duties. Yet it wasn't the chauvinist hogwash being spouted by male critics that had done for her. The main reason behind Sheila's ousting was the volume of complaints the BBC had received from women listeners. As the head of the BBC's women's section, the euphoniously named Elise Sprott explained, 'Women wrote in such large numbers that they did not want a woman announcer till, at last, we had to remove Mrs Borrett.' She added that the complaints were divided roughly equally between those who simply felt a woman's place was not on the wireless and those objecting not to the presence of a woman necessarily, but that she was a married woman when the job should have gone to a female of self-dependence who was making her own living.

Sheila staged her exit from the BBC entirely on her own terms. When she made her last announcement on the evening of Friday 25 November – 'Goodnight . . . and goodbye' – one newspaper pointed out that she had been originally due to leave a week earlier but 'exercising the prerogative of her sex, she changed her mind'.

Most of the coverage of her last broadcast lamented that, according to the schedules, Sheila's final shift was supposed to be the following day, a Saturday. But Sheila, again being Sheila, was keeping control of the narrative and had decided against going quietly. She wasn't able to broadcast on the Saturday because at that stage she'd be on her way across the Channel to Fécamp, where she'd arranged to broadcast on the fledgling commercial station Radio Normandy on the Sunday. What more perfect riposte to the BBC could there have been short of opening the door of Reith's office and mooning at him? Going immediately to the Corporation's only rival and announcing the 'frivolous' Sunday concerts and gramophone records broadcast on Reith's sacrosanct Sabbath was an elegant and classy way to make an exit.

She wasn't finished then, either. She shot back from France in order to open on the Monday night in a cabaret season at a West End restaurant a stone's throw from Broadcasting House. Sheila became the house MC, as well as entertaining audiences herself with a selection of popular songs, some with the house band and some accompanying herself on the ukulele. Not only that, a fortnight after her final broadcast she began fronting a short series of films about fashion for British Pathé.

There was no lingering resentment, however. Once her point was made, Sheila was back at Broadcasting House in January to recite the 'To fly from, need not be to hate, mankind' passage from Byron's *Childe Harold's Pilgrimage* and was soon appearing regularly in radio plays again. She took up a part-time post training students in microphone technique and also became a sought-after consultant on interior decoration.

From April 1936 the Borretts began spending long periods apart when Giles, now retired from the Navy, was appointed assistant to the BBC north region executive, based in Manchester. With her work mainly based in London – most notably fronting a broadcast production with the magnificent title *Louisa Wants a Bicycle, or, the Fight for Woman's Freedom* – Sheila spent most of her time in a flat the couple owned in Pimlico, where a BBC producer named Ian Cox took a room as a lodger.

In March 1938 Sheila was back in the divorce courts, this time as respondent on the grounds of her adultery with Ian Cox at the Pimlico flat. She didn't defend the suit, admitting that she and Cox were living together as a couple. What made things even more icky was that Cox was married, with a young child, divorcing his wife three months after the Borretts were granted theirs. Sheila married Ian Cox on 1 March 1940 at Caxton Hall in London and, despite the scandal, circumstances dictated that in the spring of 1942 she rejoin the BBC as a wartime announcer under her latest married name of Sheila Cox.

During the war Sheila presented programmes on the Home Service and the Forces programmes, reading out messages for soldiers serving abroad from their families, and in 1944 was part of a team broadcasting news updates on the D-Day landings from the basement of the Ministry of Information building, the secrecy surrounding which was so great the broadcasters were locked in the basement for three days before the event. After the war she began presenting an English magazine programme on the World Service, as well as a series called *Spotlight on Women* for the BBC's American service.

Her marriage to Cox broke down during the war, and in 1949 she was set up by a friend on a blind date in London with an American advertising executive named Gager Wasey, in town trying to drum up postwar business for his father's agency. Within three years of that first date Sheila was married again and living in New York, working in fashion for the House of Davidoff. In 1954 the couple moved from New York to Pinellas, Florida, the state where Sheila would spend the rest of her life, and it wasn't long before she was broadcasting again, co-presenting a show called *Coffee Break With Bea and Sheila* five days a week on a local station.

Other than the death of her husband in 1970 Sheila's final years seemed deeply happy ones. She began presenting a classical music show, *Words and Music*, on the Tampa station WUSF, a show that ran for 15 years until her death at the age of 81 on 30 April 1986, the result of a stroke she'd had three months earlier on the day before she was due to present her last ever *Words and Music*. She had also been in line to be inducted as a Broadcasting Pioneer by the Clearwater chapter of the American Women in Radio and Television organisation.

As Sheila Stewart Wasey she had become a notable name in postwar American broadcasting, having in addition to her WUSF show presented documentaries for the American public service broadcasting network NPR, winning a Cultural Documentary Award for a programme she made about Armistice Day and a Corporation for Public Broadcasting award for a 1982 documentary about Dylan Thomas. To this day there is a Sheila Stewart Memorial Fund for public broadcasting at the University of South Florida.

Sheila Borrett did her level best to tilt at the broadcasting patriarchy. Radio in its early days was an almost exclusively male preserve and women had a long way to catch up. Today there are more female voices on the airwaves than ever before, but we're still a long way from gender parity. It's an awful shame that Sheila Borrett's name has been largely forgotten in the canon of British radio history, for she was an extraordinary woman who, even when placed in an almost impossible position in impossible times, more than held her own. The way she handled her ousting from the BBC was as eloquent as it was marvellous. Sheila Borrett was not going to go quietly and made her noise not by getting angry or bitter, but showing the BBC precisely what they were missing. The fact that within months she was back acting in dramas and giving readings – her performance of an abridgement of *Northanger Abbey* in 1937 received particularly positive reviews, while her radio acting was fêted in print more than once by the *Observer*'s radio critic Joyce Grenfell – was perhaps the Corporation's acknowledgement that they hadn't done right by Sheila Borrett and she deserved much better.

There don't seem to be any surviving recordings of her BBC work but there is a minute or so of one of her fashion films for British Pathé available online, made just a few weeks after she'd been unseated from the BBC announcer's chair. It's a brief clip, but one in which she celebrates an all-woman fashion house with a deliciously dry wit and a slightly raised eyebrow over a steely gaze right down the camera lens that tells you that she may still be in her twenties but Sheila Borrett has taken quite enough crap from men, thank you very much. Even in that minute or so of footage you can sense a perfect distillation of her personality. Sheila Stewart Graham Whitefoord Borrett Cox Wasey was a true force of nature, a trailblazing woman broadcaster who deserves a better legacy in Britain. 'You just had your voice to get your point across,' she said late in her life. 'And I had an absolute foghorn of a voice.'

Glance at any copy of the *Radio Times* during the 1930s and it is immediately plain just how male oriented the broadcast schedules were. There's not exactly an equal gender split today, but there's probably no greater example of how radio was seen as a male preserve than a series that ran for a few years from the mid-thirties called simply *Men Talking*. And that's exactly what it was. Men, talking. 'Bright, witty, interesting talking that can hold an audience is not among the lost arts,' ran the programme's description. 'Worthwhile conversation nourished with good letter-writing in the days before the telephone and motor cars turned life into a race may be rare today, but it is hoped to show that it is not extinct.'

By 'conversation' it seems the BBC meant 'posh blokes pontificating', as a rummage through the listings reveals *Men Talking* to be nothing more than chaps with double-barrelled surnames, lots of initials or a surname for a first name, being handed the microphone and invited to address the nation on whatever subject they chose. Men like A.J.A. Symons, 'a lover of rare books and connoisseur of good food and wine' for whom 'talking is one of his recreations', or C. Whitaker-Wilson who was the Man Talking one week in November 1935 apparently on the grounds that the BBC had 'in 1924 broadcast an organ recital he gave at St John's, Regent's Park'. One week Compton Mackenzie and an Edinburgh banker friend of his were invited in to have a chat over the air, giving as their subject 'this and that', which sounds like a must listen.

129

It's frankly a wonder Sheila Borrett was allowed anywhere in Broadcasting House outside the typing pool, but there were some women who managed to snaffle positions of responsibility in the early days of the BBC. Indeed, Reith was at first surprisingly relaxed about employing women, as long as they toed the line and toddled off again as soon as they'd tossed the bridal bouquet. Sheila Borrett's initial stint at the microphone was the most public manifestation, albeit a brief one, but there were also women like Olive Shapley and Hilda Matheson ploughing distinctly feminine furrows through the early history of British radio.

The daughter of a Presbyterian minister from Putney, Hilda Matheson spent much of her early adulthood moving between Switzerland, Germany and Italy where her father sought treatment and recuperation for a series of illnesses. The family settled in Oxford in 1908 where Hilda managed to complete the history degree she'd intended to start before her father's health necessitated a peripatetic jaunt around Europe. Fluent in three languages, Matheson was recruited by MI5 during the First World War, undertaking work in army intelligence in Europe and finishing the war in Rome. In 1919 Matheson became assistant to Lady Astor, the first woman elected to the British parliament – after her election Astor credited her success to 'my zeal and her brain'. With some women now having the vote, Matheson organised cocktail parties at which important figures might meet influential women and perhaps punch a few holes in the gender divide. John Reith attended one of these events in the early 1920s and was so impressed by Matheson that he headhunted her for the BBC where she soon rose to be head of talks. It's likely that Reith was impressed as much by her contacts book as her intelligence, her connections making it possible to persuade major cultural figures to take their place in front of the microphone. These included H.G. Wells, George Bernard Shaw, John Maynard Keynes and, in what would become a life-changing moment for Matheson, Vita Sackville-West. The two became lovers within hours of Sackville-West giving a broadcast talk on 'the modern woman', commencing a relationship that lasted for many years.

Matheson's tenure coincided with the loosening of the news shackles attached to the BBC by the newspaper industry, and the BBC's first in-house news department fell under her remit. *The Week in Westminster* was a Hilda Matheson creation, devised initially to give newly enfranchised women the opportunity to find out more about the parliamentary process to which they had finally been admitted. Matheson also organised the first broadcast debate between the leaders of the three main political parties.

As the twenties progressed, however, she began to fall increasingly foul of Lord Reith, most notably when she booked talks by speakers on modern literature. Reith utterly detested most, if not all literature published post-Dickens and when Sackville-West's husband Harold Nicolson was invited to talk about the then banned works *Ulysses* and *Lady Chatterley's Lover*, the old Presbyterian's hackles shot right up through the top of his head. Nicolson was an outspoken socialist, which was bad enough, but having him talking about books that would torpedo the morals of anyone who so much as thought about them was a step too far. The ridiculous compromise brokered by Matheson was that Nicolson couldn't mention the titles of the books but he was allowed to say that it was the BBC who had prohibited him from mentioning them. Nicolson was also part of a different problem for the Director-General: there were whispers about alleged left-wing bias at the BBC even then, which prompted Reith to start leaning more heavily on Matheson and interfere with the kind of guests she was booking. By the end of 1931 Matheson decided her position had become untenable and resigned.

After a spell working as the *Observer's* radio critic, where she proved herself to be astute and perceptive about every aspect of the medium, Matheson wrote a book called *Broadcasting* that put Reith's nose even further out of joint (he described it as 'monstrous' and had the *Radio Times* publish a review so snarky it could only have come from the Reithian pen itself) and then worked on the authoritative *African Survey*, an examination of British imperial interests in sub-Saharan Africa for which she was awarded an OBE. On the outbreak of war Matheson was immediately put in charge of the Joint Broadcasting

Committee, whose role was to disseminate pro-British propaganda, including the multi-volume *Britain in Pictures*. Alas, most of the issues turned out to be posthumous as Hilda Matheson died of Graves disease after a thyroidectomy in October 1940 at the scandalously young age of 52. After her death the composer and suffragist Dame Ethel Smyth paid tribute to Matheson's 'blending of intellectual grip with what one may call the perfect manners of her soul. Her one great fault was an inability to say "no" when asked to do a service; myself, I think it was this that brought about her untimely death'.

Where Matheson's BBC career was relatively brief, by contrast Olive Shapley's time at BBC radio was long and varied, a period in which she constantly changed the perception and even sound of radio in a way still discernible today. As a producer and occasional presenter she championed the presence of real people on the radio, people whose accents were far from the received pronunciation that held British radio in its cut-glass grip for so long, and strove to tell the stories of ordinary people, men and women creating incredible documents of Britain in the thirties that, had they been written and published by George Orwell, would still be regarded as classics today. So great was her contribution to the life of the nation, particularly the northern half of the country, that in Didsbury, Manchester, where she spent the second half of her extraordinary life, there's now an Olive Shapley Avenue.

Born in Peckham, south-east London, where her father was a sanitary inspector, Shapley developed at an early age a deep-set resentment for social injustice, something she would attempt to counter in both her life and her broadcast career. She read history at St Hugh's College, Oxford, and was briefly a member of the Communist Party of Great Britain, an act of youthful radicalism that even in her sixties would provoke regular visits from MI5 – visits she came to look forward to as she and her regular visiting spook could sit down and enjoy a pot of tea and a chat.

After an unfulfilling couple of postgraduate years as a teacher, in 1934 Shapley arrived at the BBC in Manchester ready to take charge of *Children's Hour* under the Manchester operation's chief Archie Harding. Harding was a Marxist intellectual who was gathering a radically left-wing team of programme-makers safely out of the immediate orbit of John Reith, who had sent Harding to Manchester in the first place on the grounds that he wouldn't be able to do much damage there.

Children's Hour was no easy task: Shapley was plunged straight into the high-pressure demands of producing an hour of live radio every day, five days a week, including two plays. Promoted to assistant producer in 1937 she commenced making a series of groundbreaking documentaries that were unprecedented in their scope and subject matter and in the way they gave a voice to ordinary people. It was an important step in pulling radio away from the smug old boys' club exemplified by *Men Talking*. Her first was a documentary about shopping presented by the unlikely figure of future *Have a Go* quiz presenter Wilfred Pickles (one of many radio notables whose careers were given a bunt up the backside by Olive Shapley: years later she lobbied vigorously to have a young Geordie named Brian Redhead present a literary series she was producing when the BBC suspected his Geordie accent might be too much for a southern audience), after which she dropped in on the lives of canal workers, long-distance lorry drivers, the homeless and miners' wives, and documented 24 hours in the life of a grand hotel in Scarborough through the eyes of its staff. Driving around the country wrestling with the BBC's seven-ton outside broadcast van Shapley developed a rare knack for drawing out the personalities of people who were not natural broadcasters to create some of the most engaging and spontaneous radio of the age.

In 1939 Shapley produced a documentary scripted by Joan Littlewood called *The Classic Soil* which revisited Friedrich Engels' *Condition of the Working Class in England*, comparing conditions in Manchester and Salford in 1844 with those of 1939 through the voices of ordinary people telling their stories. I tend to roll my eyes when people accuse the BBC of left-wing bias, just as I do those who label it too right wing, as it usually just means they've heard something they disagree with, but anyone who heard *The Classic Soil* would have to concede the programme was pretty full-on leftie. Indeed, in the 1990s Shapley herself described the 40-minute documentary as 'probably the most unfair and biased programme ever put out by the BBC'.

In July 1939, the same month that *The Classic Soil* aired, Shapley married John Salt, the BBC's northern region programme director and, because of the prevailing opposition to women going out to work once they were married, Shapley was forced to resign her job. She was still able to work on a freelance basis, however, and in the early years of the war made several programmes documenting everyday people's experiences of the conflict, including the impressions of the countryside by urban evacuees.

In 1941 Salt was posted to New York as the BBC's deputy North American director and Shapley joined him in Manhattan, where they lived for a while in Alastair Cooke's apartment on Fifth Avenue. Shapley sent fortnightly reports from America for *Children's Hour* called *Letters from North America*, a series of dispatches that featured interviews with Eleanor Roosevelt and Paul Robeson (when Shapley invited Robeson out for dinner after the interview she was startled when he declined on the grounds that 'you and I, a coloured man, could not be seen in the same restaurant together'). Shapley also befriended Mabel Dobson, Cooke's maid, who took her into Harlem, providing her with plenty of material on the plight of New York's black population. Cooke would eventually take over Shapley's fortnightly bulletins from across the Atlantic, the slot becoming that iron horse of twentieth-century radio *Letter from America*.

The couple returned to Manchester after the war, but Salt died of stomach cancer in 1947, leaving Shapley with three children to bring up alone. Despite heartbreak and hardship – beset by depression throughout her life she suffered a serious breakdown in the US in 1942 – Shapley relocated her family to her native London and in 1949 began presenting *Woman's Hour*, giving the show a grittier, less ephemeral feel by discussing previously taboo subjects like the menopause and whether it was possible for a woman to live a happy and fulfilled life without being in a relationship with a man.

The following year Shapley began appearing on children's television, fronting *Olive Shapley Tells a Story*, and she married again, to a Manchester businessman named Christopher Gorton, moving north once more to share Gorton's enormous Didsbury mansion Rose Hill.

When her second husband died suddenly of a heart attack in 1959 Shapley endured another breakdown, but returned in 1960 to present a northern edition of *Woman's Hour* as well as a series called *The Shapley File*, which confronted pressing social issues. She retired at the turn of the seventies, and at the age of 63 travelled the world on her own before returning to Didsbury and opening up Rose Hill as a refuge for what she called 'unsupported mothers'. When social services in the area improved sufficiently that she found herself rattling around the place alone again, she exhibited the same level of concern for the less fortunate when, in the late seventies, she welcomed dozens of newly arrived and still traumatised Vietnamese boat people into Rose Hill.

When Olive Shapley died in 1999 at the age of 88 it closed a life that practically encompassed the twentieth century. There are many reasons to celebrate Shapley, but it is her documentaries of the late thirties that deserve to be acknowledged as masterpieces in their own right and fêted for changing the face of radio by giving voices to people and accents that had never before been heard on British airwaves.

For *Night Journey* for example, her documentary about the lorry drivers who ploughed up and down Britain's major roads while the rest of the nation slept, she spent two nights in a transport café recording drivers' thoughts and experiences as well as riding shotgun in an eight-ton eight-wheeler on a ten-hour overnight journey half the length of the country. For *Homeless People* she travelled to North Yorkshire and the St John of God Hospital for Incurables and Cripples, the Seamen's Institute in South Shields and a Durham institution that took in homeless youths and attempted to find them jobs. For *Miners' Wives* Shapley embedded herself in the Durham mining village of Craghead for several weeks, winning the trust and confidence of its hardy women and persuading them to unburden themselves about their lives into her microphone. She took one of the women, Mary Emmerson, to France and the town of Marles-les-Mines in the coalfields near Lille in order that she might compare and contrast the lives of the women from both communities. This was genuinely innovative, groundbreaking radio, and the undercurrent running through all her pre-war documentaries is one that remains crucial to making good radio documentaries today:

empathy. Despite coming from an entirely different background to most of her interviewees – from hardened miners to 14-year-old Archie Thompson who was her guide, fixer and companion on the Leeds and Liverpool Canal for *Canal Journey* – Olive Shapley's natural human warmth allowed her to capture the essence of northern England, with all its accents, characters and incredible human stories, and distil it into absorbing radio. Shapley recognised that the poor have as much dignity as anyone else, something rarely afforded them by the workings of state machinery.

Documentary making today is much more straightforward in terms of logistics: you go out into the field with a microphone and a digital recorder about the size of an old Sony Walkman. Olive Shapley came chugging into town with a crunching of gears at the wheel of a lorry packed with recording equipment, recording discs and cables. Everywhere she went she had to run cables from wherever it was she was conducting interviews – a kitchen table, a community hall, a hospital ward, an agricultural field – back to the disc recorders on the truck. It was incredibly hard work – where editing today involves moving slices of digital audio around a screen, Shapley was working effectively with a pile of gramophone records – yet she still made incredibly good radio. The documentary is a fine conduit for creating magical radio at its most powerful and moving. Everyone, without exception, whoever they are and wherever they're from, has a story, and Olive Shapley had a gift for finding the best human stories and turning them into great radio.

The voices of ordinary Salfordians and Mancunians in *The Classic Soil* stand out as fresh and innovative nearly 80 years later. The presenter-links and staged recreations of Engels reading his work sound stilted and false, but the women in particular still speak to us down the decades: angry, weary, dignified, intelligent and articulate, chronicling lives and issues far more thought-provoking and nuanced than two posh men sitting by a microphone discussing 'this and that' after a news bulletin read by a woman whose very gender left her ill-equipped to impart events of great import.

12

'Charlton Athletic Nil': Charlotte Green Reads the Football Results

The first thing that struck me was that I wasn't cold. As Pavlovian responses to a piece of music go it's a pretty low-wattage one, but when the rumpty-tumpty theme to *Sports Report* struck up I all but started blowing on my hands.

Normally when I'd hear that particular piece of music I'd have just got into a car after two and a half hours open to the wintry elements watching Charlton Athletic. Driver or passenger the routine was always the same: open the car door, slide into the seat, remove the match programme from my pocket and toss it onto the dashboard, commence vigorous hand-rubbing and blowing, make a numb-skulled statement of the numb-fingered obvious like, 'It. Is. *Freezing* out there,' lean towards the radio squinting against the residual light from the stadium floodlights and switch it on.

On perfect days you've see your team win and switch on the radio just at the moment the music starts. It's more likely they'll have played like Bambi on ice, been utterly walloped, you've forgotten where you've left the car, you've been strafed with sleet all afternoon and, by the time you switch on, the results are nearing the end of the Scottish lower divisions. Whatever the circumstances, the sound seems to come in like the tide in a wash of medium wave that ranges from a thin, distant voice in a storm of static to a booming, distorted basso profundo that makes the speaker mountings buzz. And it's still cold.

This time, however, when the voice on the radio said, 'Good afternoon, I'm Mark Chapman, you're listening to BBC 5 Live at five o'clock, and this is *Sports Report*,' I wasn't cold. When the music started I wasn't cold. When the day's sporting headlines came through I wasn't cold. And when Chapman ended his summary of the day's headlines and trail of imminent features and analysis with, 'That's all after the classified football results with Charlotte Green,' I wasn't cold. As the music faded to the silence that befits the solemnity of the results the reason I wasn't even remotely chilly on this particular wintry Saturday afternoon was because at that moment I was sitting in a little studio deep inside Broadcasting House right next to Charlotte Green. 'The Premier League,' she began, a huge 5 Live logo on the wall behind her as she placed a hand flat on the first piece of paper. 'Burnley one, Watford nil . . .'

The Saturday afternoon classified football results are as big a British radio institution as the shipping forecast. There's the same rhythmic intonation, the same inflections, the same measured delivery that never discriminates between giant and minnow, the poetry of lists, the litany of familiar names, some of which you'd struggle to point to on a map. I imagine most football fans would cheerily concede there are several towns and cities in these islands of which we'd never even have heard of were it not for the football results. Thanks to the weekly Saturday ritual I know the correct pronunciation of Brechin and the right syllabic emphasis for Peterborough and Kilmarnock. The Saturday football results have also helped to ensure my mental gazetteer of Britain is not so much geographical as a succession of league ladders, marked out in the season-long demarcated ghettos of the divisions.

And now here I was, chair pulled up, watching as the voice that has consistently delivered both the shipping forecast and the football results in a way that sees the words 'national' and 'treasure' coaxed together whenever her name is mentioned, gave out the bare, unvarnished numerical fallout from the dozens of gladiatorial combats that had ransacked the emotions of spectators for the previous couple of hours across the nation.

While the 5 Live output comes from the BBC's Media City in Salford, the football results are still broadcast from London. Hence where I might have been expecting a large and frantic studio packed with eager graduates and grizzled old sports hacks running around with bits of paper while producers and presenters waved their arms around and shouted a lot, instead I was in a small, quiet room swinging back and forth on an office chair between Charlotte and her producer Audrey, who collates the results from the wires and writes them onto a pre-printed sheaf of the day's fixtures. While the ballyhoo of sifting the day's stories into order of priority and lining up correspondents in draughty stadium gantries played out in the north-west an air of calm assurance prevailed in London. When I arrived the matches across the country were entering their last 15 minutes and both women watched the screens intently, occasionally passing comment, especially when it concerned their own teams (Watford for Audrey, Tottenham Hotspur for Charlotte).

In contrast to the simmering frenzy of the rest of *Sports Report* the air of unflappability seemed entirely appropriate for the secular ceremony that is The Reading of the Football Results. For one thing, they're always referred to as the *classified* football results, a term that lends unimpeachable authority: a sense that while you might have heard the scores as they came in, or scrolled through them on your phone, nothing is official until you've heard the classified recitation in the traditional order on *Sports Report*. Until then the day's scores should be taken as nothing more than hearsay.

As full-time whistles blew and the scores flashing up on the television screens began to change to the colour of permanence, Audrey set about the fixture sheets with a biro, checking, double-checking and shuffling the sheaf of paper into the correct order before handing them over the table to Charlotte, who was sitting in front of the microphone. While slightly disappointed that the official register of scores was effectively handwritten rather than, say, wheeled in on some kind of stone tablet, I thought of the thousands of people across the land who at that moment were hurrying for the exits and preparing to perform the hand-rubbing, floodlight-dazzled Saturday teatime ritual known as 'Quick, Stick on *Sports Report* for the Results'. Even in the era of the Smartphone it's comforting to think that men, women, boys and girls, generation after generation, are shushing each other in cars as the whining blowers suck in the exhaust of the vehicle in front, whose occupants are doing exactly the same, filling the car with cold, fumy air.

As Audrey slid the completed sheets one by one across the studio table where Charlotte put on her glasses, picked up a pen and regarded them like a teacher about to mark a pile of maths tests, I also found myself thinking of that football radio species rarely spotted these days and possibly close to extinction. I speak of Transistor Man.

In his heyday you imagined him working during the week for the council, in a library maybe, or doing a bit of light maintenance work with younger lads whose banter he didn't understand. He still lived with his mother, was in his forties and liked books about the Second World War and making model aeroplanes. He went to the shops for his mum and the neighbours would comment on how good he was to her. She'd cook him liver and bacon when he got in from work, and he made her a hot drink before bedtime and they were both entirely happy with that. He'd sometimes go to the pub up the road and have a half of bitter, and if the barman nodded at him in recognition he'd feel ten feet tall. His hair was shaggy and unkempt and pushed over to one side to keep it out of his eyes, over which he wore permanently misty spectacles. He was usually unshaven but never grew a beard. Transistor Man had never worked out his place in the world. He'd lie in bed at night and look at the shadow cast on his ceiling by a model spitfire caught in the streetlight outside – the plane he'd made as a boy with his dad the year before his dad had died. He imagined being a pilot. He missed his dad.

Transistor Man was a shy man but he never really felt lonely, just that life seemed to be going on somewhere else, that at some point it had got ahead of him and he'd never caught up. He was Philip Larkin's Mr Bleaney, if Mr Bleaney had never left home.

But Transistor Man loved his football and for a couple of hours most Saturday afternoons he was a giant. Transistor Man always took a radio to the match, taking up the same position on the terrace or in the stand where he was known and recognised as the central node of football news. His radio was never the small-handheld variety either, it was the one from the kitchen worktop, a big thing with a handle made by a company like Grundig; a radio so chunky he had to cradle it in his arms. Throughout the game he'd field enquiries for the latest scores. If it was a crucial end-of-season match in which promotion or relegation was at stake, people would crowd round Transistor Man for the latest updates. These were days when events at a ground at the other end of the country could have as much bearing on the tension and uncertainty as the match in front of them. Transistor Man lived for those Saturday afternoons.

Transistor Man never walked more confidently than when he left the ground at the end of the game. If you weren't travelling by car or you'd parked too far away you'd skip and scamper to catch up with Transistor Man and stay within range of his radio. Every football fan without exception – from the sharpest-dressed youngsters who go to all the away games and start all the chants, to the old boy in the flat cap and the scarf in club colours knitted by his wife decades earlier – has performed the Transistor Man Scamper, that quickening of the step to draw within earshot of the hefty old Grundig, held like a newborn by the man who for a precious, nourishing, life-affirming couple of hours on a Saturday afternoon, truly belonged.

As I watched Charlotte commencing the familiar weekly tessellation of names and numbers on which a significant proportion of the nation was hanging, I thought of the few remaining Transistor Men, right then heading for bus stops and railway stations around the country, each with their gaggle of radio disciples, each of them in that moment at the exhilarating centre of their individual universe.

'Nottingham Forest three, Bolton Wanderers two.'

James Alexander Gordon couldn't say 'Wanderers', and I thought that was fantastic. His otherwise impeccable reading of the football results was given with only a very slight hint of his Edinburgh roots, most obvious when he had to say 'Wanderers'. It wasn't just down to his Edinburgh-ness, though: he always seemed to give the word an extra syllable or two and a slightly elongated ending, so it would sound something like, 'Bolton Wanderererrrrs one, Wolverhampton Wanderererrrrs one.' That there was one slight flaw (if you can even call it a flaw) in James Alexander Gordon's delivery only made the whole thing more perfect.

There can't be many football fans under the age of 60 who remember hearing the football results before James Alexander Gordon read them. BBC radio began broadcasting football results in 1923, albeit not until after a 6.15 p.m. embargo so as not to scoop the Saturday evening sports papers. Staff announcers would read them, with the format we know today solidifying after the inception of *Sports Report* in 1948 (even if the announcers in those days said 'nought' instead of 'nil'). So synonymous did James Alexander Gordon become with this ritual 'beating of the national bounds' (as Michael Palin called it) that even the oldest football enthusiasts in the land probably think it was still him they heard back in the days of Third Lanark and beyond to Bradford Park Avenue and Clapton Orient.

It wasn't just the longevity that embedded him so firmly into the national fabric – he began in 1972 and only retired through ill health at the end of the 2012–13 football season – James Alexander Gordon defined a piece of radio like nobody else in any genre on any network. Even the *Sports Report* theme music worked for him: he was the kind of person you felt should be accompanied everywhere he went by a brass band playing a jaunty march.

It should be almost impossible to inject a distinct personality into a list of names and numbers simply by using one's voice, but James Alexander Gordon managed it, his authoritative tone suggesting he'd just taken a pipe out of his mouth before he began and would replace it between his teeth immediately after the last result. A trilby with a colourful feather in the hatband would have hung above a gabardine raincoat behind him and a tightly furled umbrella will have stood in the corner of the studio. Meanwhile, a vintage two-seater sports car was parked downstairs with a pair of leather driving gloves in the walnut-fronted glove compartment.

I've no idea what James Alexander Gordon looked like but the voice created an image in my head that I'll bet was not far off the real thing anyway. His air of authority came with a tangible warmth and playfulness; there was always a smile in his voice to soften the edges of a tone that might otherwise have sounded like he was handing you down six months without the option. Nobody with an ounce of humanity in their soul could fail to feel affection for James Alexander Gordon. It was an affection combined with a deep respect, a level of respect that dictates his full name – sounding, as it does, like a firm of Edinburgh solicitors or a Jermyn Street shirt emporium – should be used at all times. 'James Gordon' doesn't sound remotely right, simply 'James' horrifyingly over familiar, and while his friends and colleagues called him JAG, for the rest of us referring to him in that fashion would suggest an intimacy that didn't exist. No, James Alexander Gordon he was, and James Alexander Gordon he will always be.

For a man whose voice charmed a nation – and was used in Sweden by teachers of advanced English to illustrate diction and nuance – James Alexander Gordon's life was topped and tailed by crippling vocal trauma. The son of a farmer, he was adopted as a baby when his mother died in childbirth, only to contract polio when he was six months old. The condition left him paralysed and mute, with significant portions of his childhood spent in hospital and his education undertaken at special schools. He needed leg irons to walk, and speech was eventually coaxed out of him, but James Alexander Gordon had an active mind and had been reading voraciously from a very early age. In addition, he spent long hours listening to the radio, including Saturday teatimes sitting with his father, watching him become infuriated as he tried to check his pools coupons, only for the results to be read out too quickly.

A gifted musician, James Alexander Gordon embarked on a career in music publishing, licensing the work of bandleaders like Bert Kaempfert and James Last, which is absolutely the kind of music you'd expect James Alexander Gordon to have had in his record collection. There would be Beethoven symphonies and Chopin piano sonatas all right, but some days he'd arrive home from reading the results and only *A Swingin' Safari* would do.

As he entered his mid-thirties there was no sign that, even though it was an ambition he'd nursed as a boy, broadcasting would ever be a career option. Then around 1971 he was asked to write and record a voiceover for a television documentary about one of the artists he represented, and James Alexander Gordon's life took an unexpected turn. It was four years after the BBC Home Service and Light Programme had been abolished in favour of Radios 1, 2, 3 and 4, and as an ongoing extension to this long overdue modernisation the Corporation was looking for broadcasters with a more regional variation in their voices than the narrow Oxford and Cambridge recruiting seams they'd mined for the previous five decades.

James Alexander Gordon was invited to become an announcer, reading the news, the weather forecast and the shipping forecast, and after reading the four o'clock news one Saturday afternoon in 1972 was asked if he'd pop over to the sports department and read the football results on *Sports Report*. He filled in for a fortnight but made such an impression in the weeks that followed the BBC was flooded with pleas for his return. In the ensuing 40 years he missed only one Saturday afternoon shift, when an accident on the M4 prevented him from reaching Broadcasting House from his Berkshire home in time.

James Alexander Gordon's definitive style, the rising and falling tones depending on victory or defeat, grew out of his compassionate nature and musical background. If a team lost, he said, he felt sorry for their supporters and so used a downward inflection while indicating delight for the winners with a suitably uplifting tone. James Alexander Gordon had no bias towards any club as he didn't support one – before the whole *Sports Report* staff attended a sixtieth birthday bash at an Arsenal versus Manchester United game in 2008 the last match James Alexander Gordon had attended in person was a Falkirk v Celtic tie in 1948 when he was 12 years old.

In a cruel stab of fate his childhood muteness returned to bracket his life at the end when treatment for the throat cancer that would eventually kill him necessitated the removal of his larynx. A short time later, on 18 August 2014, the extraordinary voice that had provided the soundtrack to countless exultant hopes and shattered dreams was stilled for ever.

When James Alexander Gordon stopped reading the football results there was almost a sense of the ravens leaving the Tower of London: if the country wasn't about to fall as a consequence then surely the game itself might? Luckily, the BBC made exactly the right choice of replacement, the woman who had by now reached the League One results next to me. I already knew what was coming.

'Charlton Athletic nil, Portsmouth one.'

When you're in the car and your own team's result goes through it's impossible not to react. For a start, it's often the only result you already know for certain. There's a triumphant cheer if you've won but if the result hasn't gone your way there's no single prescribed reaction. There could just be a tut from the back seat, especially if it's been a heavy, dispiriting loss with which everyone in the car has already come to terms. For closer results the responses will be more assertive. 'Get out of it,' someone will say. 'Shut up.' 'Turn it off.' 'Pricks.' 'Charlton Athletic nil, referee one, more like.' For a draw you're most likely to emit a subdued non-committal cheer, as if your team just being mentioned on the radio is an achievement worth marking. Whatever the outcome of your game, there is never, ever, complete silence when it's announced on the radio.

'Accrington Stanley against Swindon Town, match postponed.'

Games that have fallen victim to the weather before kick-off are never cancelled, annulled or deferred, they're always postponed. Similarly, matches that kick off at 5.30 because they're being shown live on subscription television are still announced as a 'late kick-off', as if we're all pretending the away team's coach had broken down or someone had forgotten the key to the dressing rooms. Some television results services go with, 'result not yet in', a clunky abomination that abrogates any sense of tradition or romance, but by giving out the absolute minimum of necessary information the results on *Sports Report* pitch it exactly right.

The football results may sound utilitarian and minimalist in principle but they contain a clutch of rituals and micro-traditions, quirks and tropes so entrenched and long-established that they're almost folkloric. The reading always commences with the top division in England and works its way down; then it's the same for the Scottish leagues and, a more recent introduction, the Welsh Premier League. It's the absolute opposite of showbiz, where you'd open with a wannabe or neverwas and build up towards the star turn. The football results open with the headliners and finish with the also-rans, and we still all stay listening right to the end.

Whatever the level there's always a sense of excitement and mystery when a fixture is announced appended by the phrase 'match abandoned', a suitably enigmatic summary of a rare occurrence suggesting biblical weather at best or a grave injury to a player at worst. It's something out of the ordinary, a story to relate in the pub later or ring your dad about if you were there, or to wonder about, searching Twitter or leafing through the round-up columns of the Sunday papers the next morning.

'Newtown against Cardiff Metropolitan, match postponed.'

Despite the hugely appealing fact the classifieds don't build to any kind of climax, finishing the results on a postponement felt a little flaccid. The classifieds are radio sport at its most raw – the kind of immediacy on which radio broadcasting was first built – an afternoon of drama and seesawing emotion spun in a centrifuge from which only the most basic building blocks of football life are retained, the nitty-gritty. The rest of *Sports Report* is there to commence putting the flesh on the bones of the scores, but for Charlotte Green, the bringer of tidings good, bad and indifferent, the important work has been done.

She put away her spectacles, rolled back her chair, picked up her handbag, wound a scarf around her neck and suggested we adjourn to the bar of a nearby hotel for a glass of wine. Walking out of Broadcasting House we talked about how matches seem to be finishing later these days – where once the final scores would begin filtering through from 4.45, today most games seem to be ending closer to 5 p.m. Half-time must be getting longer, we surmised.

We took the lift to the top floor of a hotel a stone's throw from Broadcasting House and emerged into a quiet, dimly lit bar with an extraordinary panoramic view across London. We sat down and ordered a couple of glasses of rosé, and I enquired as to how on earth Charlotte Green had come to be reading the football results.

'I've always been mad on football,' she told me. 'My father would take me with him to football and rugby matches from when I was very young, and I can remember from the time I was no more than six or seven years old that I'd sit at home with the newspaper and read the football results out loud. And now, in one of those strange occasions when life seems to turn full circle, here I am, 50 years later doing it for real.

'During the 2012–13 football season it became clear that poor JAG really was very ill, and he told the BBC he wouldn't be able to continue beyond the end of that campaign. I've always been a bit of a football bore, discussing the previous day's games in the newsroom, and one day, not long after I'd left Radio 4, a friend of mine rang me up and asked whether I'd like to read the football results. I thought he was pulling my leg. Apparently someone had heard me say during a feature on *Newsnight* about my leaving the BBC that one of my unfulfilled ambitions was that I'd love to have read the football results. This filtered back to Richard Burgess, who was then the head of sport. He got in touch through this friend of mine and asked if I'd go and see him. I met him for a coffee, we talked non-stop about football for about 35 minutes, so he knew I knew my stuff. We went into a workshop, he gave me a few scripts to try, and it all went absolutely fine. In terms of broadcasting I think I'm at my happiest on these Saturdays. It really was a dream come true to be given the job, and here we are five years later and I still love it. JAG wrote me a lovely email when he heard I was going to take over: "Charlotte, just make it your own." And that's what I've done.'

So used have we become to hearing the football results being read in a certain way, on television and radio, commercial and BBC, that it's easy to forget that the broad inflections shared by every reader stem largely from James Alexander Gordon. The rise and fall in pitch of a draw, the crescendo of faint surprise at the peak of an away win, the falling cadence of a home victory: like many of radio's core aspects (news headlines, the shipping forecast), someone did it first and laid down for future generations the accepted formats and inflections.

Appointing Charlotte to the job was a bit of a masterstroke. James Alexander Gordon was probably the hardest radio act to follow since Roy Plomley, and by hiring an already well-known and trusted radio voice, yet one that sounded completely different to her predecessor, BBC radio made one of its best decisions of recent years.

'It was no good just being a pale imitation of JAG,' said Charlotte. 'I have to be true to myself and I have given the style he handed down certain shades and colours. It's a lovely, iconic little five minutes that is like the shipping forecast in many ways. You have that measured way of reading, and there are some wonderful names, particularly in the early rounds of the Scottish Cup. Teams like Buckie Thistle, Forres Mechanics, Civil Service Strollers and Gala Fairydean – there is a poetry to the football results. For me as a fan it's thrilling sitting there as the results come in, particularly on a big FA Cup day. People aren't as excited by the Cup any more, which is a great pity as it really meant something when I was growing up. Third round day was always a huge occasion, with the big clubs arriving in the same round as the minnows, some of whom had fought their way through from the qualifying rounds.'

As we looked across the lights of the city, with Broadcasting House just visible behind the church of All Souls, Langham Place, which we'd passed on the way in, lights blazing, a choral recital in full flow, the conversation turned, as it always does with radio people, to early memories.

'I didn't understand a word of it, but I must have been around four years old and hearing comedy series like *Beyond Our Ken*,' Charlotte told me. 'My sister told me a few years ago that when *Listen with Mother* was on I'd go and stand right by the radio, as if Daphne

Oxenford was actually inside it. Recently a friend of mine told me her little boy would do exactly the same when I was on, even walking up to the radio and saying hello, thinking I actually lived inside it.

'I think we were very much a family that was keen on the radio. We owned a television but it was rarely on: we listened to the radio much more. I think when a radio programme is good it really engages you on every level in a way that television can as well – there are some brilliant television programmes, I'm not dismissing it by any means – but sometimes while watching television I'm not thinking very hard, I'm just looking at the pictures and absorbing the words. The best radio, and it happens a lot more than it does with television I find, actively engages the listener with the programme. You use your mind in a different, more proactive way than you do when watching television, and I think that's what appeals to me most about radio.'

After leaving school Charlotte took a degree in English and American Literature at the University of Kent, gaining first class honours despite the distractions of the medium in which she would make her career.

'I did a lot of work with the university station,' she recalled. 'They have a fantastic, deeply impressive community radio station there now, one that broadcasts across Kent. They were kind enough to bestow on me an honorary doctorate a few years ago, and on the day I went to receive it they also asked if I'd open their new media centre. The difference in the facilities from when I was there is incredible: we had these tiny studios with egg boxes on the walls – now they have state of the art facilities for both television and radio.'

Having gained a taste for radio at university, egg boxes notwithstanding, Charlotte had her head turned from her previous ambitions that lay in the direction of an acting career. Before her finals she applied for the same route by which Corrie Corfield had arrived at the BBC, the studio managers' training course.

'One of my parents' neighbours was a producer on *Woman's Hour* and she told me it was a good way into the industry because you learn every aspect of putting a radio programme together, which sounded interesting, so I filled in the application and posted it off. A letter arrived on the morning I did my final paper but I couldn't bring myself to open it until I'd finished the exam. When I did I learned that I'd been accepted.'

Once she'd commenced the course it was clear that Charlotte had a perfect voice for radio and it wasn't long before she was asked if she'd ever thought about presenting.

'I hadn't, really. I wasn't terribly ambitious, I was just having fun. Being a studio manager at the BBC in those days was a little bit like being back at university – it was the social aspect that I enjoyed as much as the work itself. But I thought, "Well, why not?" so applied for an announcers' attachment.'

During her six-month stint Charlotte was asked to provide readings for a programme called *News Stand*, a light-hearted 15-minute look at the papers that also encompassed reviews of publications like *Psychic News* and the *Meat Trades Journal*. 'I had to do a Margaret Thatcher impersonation at one point, which got onto *Pick of the Week*,' she recalled, 'and I thought, "I love this, I'm really enjoying it."'

A vacancy arose at the end of Charlotte's secondment and she joined the presenting team. This was the end of the 1970s, however, and women announcers were still frustratingly rare. The old guard still ruled the roost and in some aspects the Corporation was being run in the 1980s as if it were still the 1930s.

'When I started I was under a presentation editor called Jim Black who was a slightly strange man, as tough as old boots,' said Charlotte. 'He had very high standards and was difficult to like, to be honest. Indeed only a couple of years before I started in his department he'd been quoted as saying, "If a woman could read the news as well as a man there would be nothing to stop her doing it, but I have never found one who could. A news announcer needs to have authority, consistency and reliability. Women may have one or two of these qualities, but never all three."'

I let out a kind of strangled whimper and Charlotte rocked back in her seat, laughing.

'I know! That always draws a gasp when I tell people. I mean, it was *tremendously* patronising. Luckily, when I started, my mentor was Laurie Macmillan, a huge character, very funny, a woman announcer who'd been doing the job under Jim since 1975, and she gave him as good as she got. I doff my hat to Laurie. She taught me an awful lot about what it takes to be a good announcer and newsreader, and she also made it fun. In those days you cut your teeth on schools radio, where there was this dreadful programme called *Music and Movement*. One day, when I'd not been in presentation for very long, Laurie and I were in the studio when we heard this woman say with an immaculate cut-glass accent, "And, boys, I want you to throw your balls in the air, clap your hands, and catch them again." I caught Laurie's eye, we both fell around laughing, and I knew we were going to be great friends from then on.'

Laurie Macmillan died from breast cancer in 2001 at the age of 54.

'She is still hugely missed,' said Charlotte, looking out wistfully at the lights of the city. 'She was a kingpin at that time, the first woman Jim had ever taken on, and she was very good at curbing his excesses.'

In the years that followed Charlotte's voice became one of the most recognisable in the country, almost the definitive sound of the BBC and Radio 4 in particular. Yet as we'd walked out of Broadcasting House and through some of the Saturday afternoon crowds filtering away from Oxford Street nobody had so much as done a double take, nor was there even the kind of look that said, 'I think I know you from somewhere.'

'I like the anonymity of radio, because I'm quite private and a little bit shy,' Charlotte told me, 'but people have recognised my voice a number of times. I once went into a jeweller's and asked the man behind the counter if he might repair my watch and he said, "You're Charlotte Green." He always had Radio 4 on in his workshop when he was repairing watches and had recognised my voice as soon as I opened my mouth. But that's nice, I quite like that because then you can have a chat.'

Tentatively I wafted the phrase 'national treasure' across the table and there was a barely discernible wrinkle of the nose.

'It is flattering but all I've ever wanted is to do my job as well as I possibly can. I'm a bit of a perfectionist and set myself high standards, so it pleases me enormously that people generally seem to like what I do. But I think calling me a national treasure is over the top and I'm a bit uncomfortable with it. There are far more deserving recipients of that epithet than me. I just happen to do a job that's public. There are people out there like inspirational teachers and amazing doctors and nurses who ought to be fêted ahead of someone like me. I'm always very happy when people say they miss hearing me on Radio 4, and how much they enjoyed my newsreading. I love that. But national treasure? Not so much.'

Like James Alexander Gordon with the football results it's hard to think of a time when Charlotte Green wasn't reading the news, but on 18 January 2013 at six o'clock in the evening she read her final bulletin. The opportunity for voluntary redundancy had arisen and, along with her colleague Harriet Cass, Charlotte decided it was too good an opportunity to turn down.

'As a newsreader the highlight of my years at Radio 4 was the fall of the Berlin Wall,' she told me. 'My sister lived in Germany, and only eight months before it came down we'd gone into East Berlin. I almost didn't get in when they saw I worked for the BBC, but fortunately my sister is a fluent German speaker who works for the Foreign Office so she managed to smooth things over. As we passed through the checkpoint I remember saying to her, "This is never going to come down in our lifetime." Yet there I was just eight months later reading out its demise on the news. Looking back I think it was the only positive news story I ever read, everything else was all death and disaster, and this was an extraordinarily positive thing.'

Probably the most famous (or infamous) moment of Charlotte's newsreading career came on the *Today* programme in the spring of 2008 when, after a story that included a recording of the world's oldest sound recording – a hissy, droney, wobbly rendition of 'Au Clair de la Lune' from 1860, she dissolved into uncontrollable laughter while having to read, of all things, an announcement of the death of the American screenwriter Abby Mann. What set her off was the *Today* presenter James Naughtie commenting off air that the recording sounded like a bee trapped in a bottle. The subsequent giggle-fest made the newspapers and even an episode of *QI*.

'I was worried I'd get into serious trouble over that but it was fine. I'd no idea the recording was coming up: we were actually going to come out early from the bulletin but they shoved it in front of me at the last minute with another story behind it about the death of Abby Mann. When the clip began we all started laughing in the studio and didn't realise how short it was. Then the green light came on and I had to read, sight unseen, this story about a man who'd just died, and I just couldn't stop giggling. We're a decade on from it now and I can still remember the feeling of the back of my neck prickling. Of course, once you start giggling it's impossible to stop, especially when you've got Jim and Ed Stourton shaking and snorting across the table. Probably until the day I die that will always be associated with me. If you Google me it's always "giggles" that comes up first.

'It was quite a while before I could bring myself to listen to that clip but now I can, and it's become a happy memory because I loved working on the *Today* programme. You had the best seat in the house, sitting there while John Humphrys held some politician to account. It was great theatre because you could see their body language; that they didn't want to tell the truth, and they'd squirm as he nailed them, which was great to watch. There was something about working with Jim Naughtie though, he'd always set me off. There was one occasion where I had to read a news story about the chief of defence staff in Papua New Guinea who was Major General Jack Tuat, pronounced "Twat". Immediately after that there was a story about a sperm whale being stuck in the Firth of Forth.'

When we met Charlotte was just coming to the end of a five-year stint presenting her own programme on Classic FM for which she'd interview well-known music performers and enthusiasts about their taste in classical music. Her guests had included Stephen Fry, Judi Dench and Idris Elba.

'The five years have gone really quickly but I want to pursue other things and I have a few irons in the fire back at the BBC,' she said, 'which is, I think, my spiritual home. When I left Radio 4 I wanted to take a history of art course but Classic FM came along literally the day after it was announced I was leaving the BBC and I thought it was too good an opportunity to pass up. I really want to do the course now, though, before I get too decrepit. I've loved the interviewing aspect of the programme: I'm fascinated by very talented people and some of them have been extraordinary, opening up about insecurities and fears, thinking they weren't good enough. I interviewed Murray Perahia who's an amazing pianist but really quite beset by insecurities. I loved that aspect, but I wanted to spread my wings a bit. Also the programme was becoming almost full time, which was not really an avenue I wanted to pursue.'

While it's not exactly a crash-bang-wallop who's-on-line-one? kind of affair, Classic FM is still a commercial enterprise and one very different in style from the BBC. Indeed, the station has been at least partly responsible for a shakeup at Radio 3 that, while not exactly dragging the highbrow station kicking and screaming into the modern world, has at least hauled it as far as the second half of the twentieth century. For Charlotte, the differences in style were pronounced.

'Their house style is not the natural fit for me,' she admitted. 'This sounds slightly critical, and I don't mean it to be, but it's quite gushing: everything is fantastic, everybody is a wonderful pianist or a fabulous violinist, whereas I think I'm a natural sceptic and found that constant enthusiasm quite difficult. There was also the commercial aspect of bigging up the station all the time, mentioning its name every time you opened up the microphone – that was quite a big chasm to jump across from what I was used to. I think I found my niche eventually but it took a while to get used to the difference. I wouldn't naturally gush about anything, so it made me cringe a little because it felt so false. I've enjoyed Classic FM very much, though, and it's opened up new avenues for me. I'd only done interviewing on a very small, low-key level before, and I learned a great deal very quickly, for which I'm very grateful. For one thing, I was able to interview most of my heroes and heroines: the only one I couldn't get was Simon Russell Beale, as he was always too busy. Rather frivolously I suggested Harry Kane but that didn't quite fit the arts and culture brief, alas.'

She drained her glass and looked out towards Langham Place.

'I did enjoy it,' she said, 'but my heart belongs over there in Broadcasting House. I'm steeped in it. Mine's been a kind of BBC life, in a way.'

We parted outside the hotel and I headed towards the underground. As I descended the steps into Oxford Circus station I became suddenly aware of the subterranean network circulating beneath the city. Mingling with shoppers, weekend workers leaving their shifts and people heading out for the evening were the folk travelling home from the matches whose outcomes Charlotte had broadcast to the nation – fans of all sorts of clubs winding their way around the system digesting the matches they'd just seen. They were above ground too: away fans pulling out of London's termini on long trips, some of whom wouldn't arrive home until the witching hour; home supporters looking forward to being in the warm and having a bit of dinner before they headed out again. There were people in cars, probably just about free of the gridlocked traffic coming away from the football grounds by then, making up that network of national ritual defined by the five o'clock football results on *Sports Report*. On the Victoria Line I sat next to a man with his match programme open at the league table, placing his finger next to his team and doing the mental arithmetic that would tell him how his club's and their rivals' results would affect their places in the sporting national grid. I want to nudge him and tell him that I've just sat next to Charlotte Green while she read the football results, but I don't because a) it's London where no strangers interact on the tube under any circumstances and b) the football results on *Sports Report* don't really belong in the physical realm. They are of the ether – and above our heads and above the ground, into the clear starry night sky above the bright lights of the city, the joys and disappointments of the afternoon's sport were still loose in the atmosphere where dissipating fortunes and emotions mix with medium wave.

13

'Don't Swear': the First Commentator

When Captain H.B.T. 'Teddy' Wakelam arrived at Twickenham on Saturday 15 January 1927 he was directed to a flimsy-looking shed built on scaffolding in the south-west corner of the ground. He could see it swaying slightly in the wind. Having clambered up the ladder and stepped inside he found a floor strewn with cables. A technician had pinned a handwritten sign in red ink above a window through which he could see the pitch. It had 'DON'T SWEAR!' written on it. A microphone sat on the table under it. Wakelam had begun the week just like he had any other, working quietly at his desk at the head office of a building contractor, going through piles of paperwork. And now here he was at Twickenham, about to broadcast a live description of the international between England and Wales, and in the process become Britain's first sports commentator.

With less than two hours to go before kick-off the prospects for some enlivened sports commentary weren't particularly auspicious. Teddy Wakelam had never been on the wireless before. His summariser, Charles Lapworth, was a screenwriter not long returned from years in the USA who knew nothing about rugby. Then Lance Sieveking, the man in charge of the outside broadcast, appeared in the doorway. With a blind man.

These days sports commentators are broadcast professionals who have succeeded in a fiercely competitive field, many of the younger ones having studied their art at university and worked their way up through local radio. The events that led to Teddy Wakelam taking the microphone weren't quite so conventional.

The only advantage Wakelam had in the world of broadcasting was that he came from the public school tradition that had nurtured the nation's early broadcasters. On leaving Marlborough he went up to Pembroke College, Cambridge, to study medicine only to discover almost immediately that he fainted at the sight of blood so switched to history instead. An all-round sportsman, Wakelam played cricket, tennis and rugby to a high standard, turning out for Harlequins. He joined the Royal Fusiliers at the outbreak of the First World War, rose to the rank of captain in 1917, and joined the building firm Mowlem after resigning his commission in 1921. He captained Harlequins after the war until a 1924 knee injury reduced his opportunities to turn out regularly, but he still played cricket to a decent club standard and kept up his involvement with tennis by becoming an umpire at Wimbledon.

His job in construction contracts was a steady one and Mowlem was a large enough company that his career prospects looked sound until, on the morning of 10 January 1927, the 33-year-old was summoned from his desk to take a telephone call. A voice at the other end asked if he was the same Wakelam who had played rugby for the Harlequins. Receiving an answer in the affirmative, the voice identified itself as belonging to Lance Sieveking, head of topical talks and news at the British Broadcasting Corporation. He wanted to see Wakelam at once, he said, concerning an urgent matter.

Intrigued, Wakelam went to meet the BBC man who informed him that the newly inaugurated British Broadcasting Corporation had decided to widen its remit beyond dance bands and improving lectures to broadcast live commentaries of sporting fixtures, starting with the England v Wales rugby international a few days hence. The only problem was that while they'd assembled a red-hot technical staff to set up the broadcast, they'd realised they had nobody to do the actual commentating. What they needed, and urgently, was a man who knew his rugby and might be relied upon to give an accurate description of the game's events as they happened.

Wakelam's name had come up in the course of the conversation. Might he be interested?

He thought of the teetering pile of papers on his desk back at the office, the one that never seemed to diminish. 'I'll certainly give it a go, old man,' he said. As Wakelam related later in his autobiography, Sieveking had, after all, told him that American stations broadcast live descriptions of baseball, and if an American could do it how difficult could it really be? The two men arranged to meet at the Guy's Hospital sports ground in south London the following morning, Tuesday, where they would set up a microphone and see how Captain Teddy fared commentating on a match taking place there between two teams of hospital staff.

When Wakelam arrived at the Guy's ground the next day he found an anxious Sieveking and no BBC engineers. A mix-up in the arrangements meant that while the producer and commentator were stamping their feet against the cold at Honor Oak Park, the engineers were parking their van five miles away in Greenwich Park. By the time Wakelam and Sieveking's taxi screeched to a halt in Greenwich they found the technical team standing by their van with hands in pockets and the equipment all packed away. When they'd arrived they'd assumed a rugby match between the Royal Naval College and Blackheath Wednesday about to kick off was the one they wanted and began dragging their boxes of gear across the sward. Just then a park keeper appeared, took one look at the technical jiggery-pokery being hauled across his royally appointed turf and told them in no uncertain terms to clear off. A new plan was made for the next day: Wakelam knew of a schools match taking place at the Old Deer Park in Richmond. Would that do? With the big match at Twickenham only days away Sieveking said it would have to.

Things went a little more smoothly in Richmond, where Wakelam found himself in a blustery park sitting at a table at the side of the pitch, steadying a microphone with one hand and trying to keep his hat from blowing away with the other, while describing 30 schoolboys charging around a patch of mud. Sieveking reported back to Broadcasting House that his man had done perfectly well, but then, with barely two days to go, he'd have been obliged to say that even if Wakelam had inserted an inventive swearword into every sentence and spent the entire commentary chewing toffees.

Before that chilly Saturday in 1927 the closest British radio had come to a live sports broadcast was five years earlier when Marconi's 2LO station had related live bulletins from Olympia, where the French boxer Georges Carpentier was taking on Kid Lewis. It wasn't a live commentary as such – the company's W.R. Southey had merely given out occasional brief updates from a fight that had lasted barely two minutes before Lewis was on his back with cartoon birds twittering around his head, an anvil-fisted right-hander from the Frenchman having landed on the point of his jaw.

Standing in that rickety shed on stilts, watching the first of the day's 52,000 spectators drifting into the enclosures, the building-trade contracts facilitator who'd played a bit of rugger in his day realised the enormity of his task. He was about to commentate on one of the biggest sporting occasions in the calendar, live, as it happened, for the full 80 minutes of play plus preview and analysis, with only a single chilly afternoon in a windy park with a bunch of kids as a practice run.

It's fair to say Wakelam must have felt a little apprehensive as he picked his way through a tranche of technicians and BBC officials, of whom Sieveking was the only familiar face, to reach his seat. He was even a little relieved to see the notice reminding him not to swear. 'Not that I any more than the rest of my fellow men am addicted to the use of strong language,' he said later, 'but sometimes, watching rugger, one is apt to get carried away.'

Despite the frantic, seat-of-the-pants manner in which the broadcast had been put together Sieveking and his colleagues had pulled off some fiercely inventive thinking ahead of this wild shot in the dark. By today's standards the whole thing sounds like a readymade disaster. We're used to professional commentators with a summariser sitting alongside them, someone with the insight and experience of a participant who can interpret what's happening in a way that enhances the work of the person describing the match itself.

Teddy Wakelam was a commentator who'd captained Harlequins. His sidekick Charles Lapworth, meanwhile, had briefly edited the *Daily Herald*, worked in Hollywood with Charlie Chaplin then returned to Britain to turn out screenplays for Gainsborough Pictures – but he'd never played rugby in his life. He didn't even know the rules: during the game he would ask Wakelam questions such as, 'Do they always use an oval ball?' Daft though such enquiries sound to a rugby fan, Wakelam soon realised that his audience was much wider than the kind of rotund, beery coves shouting, 'Play up, 'Quins!' from the touchline, with whom he was used to discussing the game. Lapworth was merely asking questions that listeners at home might also be posing and it was his job to explain what was going on to the novice as much as the enthusiast.

Then there was the blind man Sieveking was leading gingerly over the cables to a specially prepared seat. The producer had arranged for a patient from St Dunstan's residential hospital in Regent's Park, established for men who had lost their sight during the war, to join Wakelam and Lapworth in the commentary box for the afternoon. The man was a former rugby player who was now completely blind, and Sieveking had told the fledgling commentator not to worry about the audience listening at home but instead to address his commentary to the man from St Dunstan's. If Wakelam could describe the events effectively to him, he surmised, then the thousands listening on their wireless sets would be doing just fine too.

The clock ticked down until, with ten minutes to go before kick-off, the London studio handed over to Wakelam and Britain's first live sports commentary was underway.

'Straightaway I forgot all my nervousness and stage fright,' Wakelam recalled later, adding that he remembered nothing at all of the game or anything that he'd said. Relieved after handing back to the London studio for a couple of minutes' respite at half-time, as he turned away from the microphone the nation heard Wakelam ask, 'Now, what about a beer?'

Whether he was handed one or not, the second half passed without commentarial mishap as England won the match 11–9.

Not only had the debutant on the mic pulled it off, *The Times'* chief sportswriter Bernard Darwin heralded Wakelam's commentary as 'extremely successful', noting that he'd chosen the tone of 'an acutely interested spectator' who 'made the microphone seem a very human instrument'.

The blind ex-serviceman told Wakelam, 'It seemed that you were talking to me, and to me only, just as if you were telling me something in an ordinary conversation.'

The only dissenting voice came in a letter from the wife of a former Welsh rugby international who labelled Wakelam 'a menace' because her husband had been acting out the plays as he described them and demolished a number of items of furniture in the process.

So well had he done, in fact, that just seven days later Wakelam was dispatched to Highbury Stadium to commentate on a football match between Arsenal and Sheffield United. The match would be carried live and the BBC had also arranged to broadcast half an hour of 'community singing' from the spectators led by a brass band, the very thought of which sounds utterly horrendous. It was a freezing cold day and the match was in grave doubt until half an hour before kick-off because a layer of frozen water lay on top of the pitch. A smothering of salt and sand made it playable in the view of the referee, but the delay meant that a crowd already diminished by the awful weather had only started to filter in when Broadcasting House handed over to Wakelam, leaving him to fill in desperately until the singing started. By the time the game kicked off it had grown so dark and wet it was difficult to see clearly what was going on and who was who. The match was a pretty awful one, with the players slipping and sliding on the surface and the ball sticking in puddles. The wet conditions also soaked the ball until it grew so heavy it would have made a passable cannonball, with some of the players left visibly groggy after trying to head the thing. All in all it was a commentator's nightmare, with both goals in a 1–1 draw coming in the final ten minutes.

The damp occasion at Highbury didn't seem to dampen Wakelam's enthusiasm for commentary, however. Indeed, just six weeks after that first Twickenham commentary he was in Cardiff for the rugby international between Wales and France, during which he gave part of his commentary in French for the benefit of the visitors, something that should absolutely be reintroduced today. I, for one, would definitely tune in to hear Alan Green commentate on a Champions League match between a British club and, say, FK Qarabağ if he had to do ten minutes of each half in Azerbaijani.

By the end of the month the BBC was already branching out its coverage: on 29 January while Wakelam commentated on the FA Cup tie between Wolverhampton Wanderers and Nottingham Forest for the London and Daventry stations, other BBC regions carried commentaries on matches at Sheffield Wednesday, Port Vale and the game between Corinthians and Newcastle United. That year Wakelam also went on to commentate at Wimbledon and on county cricket, making a versatile start to a broadcasting career that would keep him at the BBC until the outbreak of the Second World War. His spell behind the mic also included supplying the first televised cricket commentary in 1938 and accidentally setting fire to his trousers at a crucial moment of a Wimbledon semi-final.

In coping so unflappably when dunked into the unknown waters of sports broadcasting Teddy Wakelam must have had some amount of sang-froid about him, but then he had been mentioned in dispatches twice and wounded during the First World War, so talking about 30 mud-caked men wrestling each other over a ball might not have been as daunting for him as it might have been for you or me. Indeed, before long, Wakelam was even able to formulate four basic principles of sports commentary that still apply today: be natural, be clear, be fair and be friendly.

In fact, those are four principles that should not only apply to sports commentary but broadcasting in general. They could also apply to life itself, but then, of course, most radio *is* life itself.

14

'And the Sun Shines Now': the Greatest Commentator

One of the first records I ever owned was an LP called *Memorable Cup Finals*. It combined recordings of BBC radio commentary from notable FA Cup finals followed by the winning club's song. Some of the commentary–song combinations were a little incongruous, such as the thrilling climax to the 1953 'Matthews final' between Blackpool and Bolton Wanderers rendered by Raymond Glendenning followed by a musical love letter to the seaside club performed by the Nolan Sisters.

Side A began with a recording of the crowd at the 1957 showpiece singing the traditional Cup final hymn 'Abide With Me' followed by a Blackburn Rovers team singing their way uncertainly through a tribute to their own prowess – a recording you could sense was embarrassing for all involved. The sleeve mentioned the 1928 FA Cup final of which there was no recording, but the other featured matches included radio highlights pulled from the archives and released as part of the 1981 commemorations of the hundredth FA Cup final. From Alan Clarke shouting 'Bobby Smith has scored for Spurs' at the 1961 final as if he was shouting the news through an upper-deck window of the bus, to a young Brian Moore's excitable description of Derek Temple streaking through to score Everton's winner in 1966 I played that record so often that I can still recite the commentaries word for word to this day.

That album was a masterclass in sports broadcasting. The best commentators make the biggest occasions memorable by not just describing what's happening but enhancing it, and on that 12 inches of vinyl were some solid gold examples of the art. The last piece of commentary on the record, of the final moments of the 1979 showpiece between Arsenal and Manchester United, was always my favourite. With four minutes remaining Arsenal were 2–0 ahead and the match looked to be drifting towards its conclusion in warm late-afternoon sunshine. And then everything changed.

'Jordan . . . Gordon McQueen! 2–1. Lovely work by Gordon McQueen.'

The voice belonged to the Welsh commentator Peter Jones, the greatest football commentator in radio history. He'd seen everything in his time, and described it. The fact the goal scored by Manchester United's Scottish international defender was noted as nothing more than 'lovely work' demonstrated how the entire watching audience knew it was a late consolation: the outcome of the match was not in doubt and the Cup was going to north London.

But then. 'McIlroy is into the area,' said Jones. Then, his voice more urgent, rising in pitch, 'He's still there, he must score . . .'

And then, with a voice suddenly rasping with excitement, 'McIlroy has scored for Manchester United!'

A breath.

'It's two-two!'

Jones's summariser Denis Law took over, babbling for a few seconds over the din of the crowd, before Jones had to break in again.

'Here comes Brady, though, for Arsenal,' he said, sounding almost drained. 'Perhaps they want to finish this off before extra-time. The ball floats high across the area . . .'

His voice rose exactly in tandem with the roar of the crowd ahead of an explosion of sound as Jones's voice soared over it, quivering with the thrill of the moment.

'The shot comes in,' he said with an air of finality, sounding almost done in by what he's witnessing. 'It's there. Alan Sunderland.'

He pauses for breath, and faintly in the background you can hear Law emit a disbelieving 'Jeez', before Jones confirms what he'd seen.

'It's 3–2 for Arsenal and I do not believe it.'

A heaving breath, almost a sob.

'I swear I do not believe it.'

Even now, replaying that commentary in my head from memory nearly 40 years after the events it described, the hairs stand up on my forearms and my breath catches slightly in my throat. I'm no fan of either Arsenal or Manchester United, but Peter Jones's commentary of that extraordinary climax encapsulates perfectly one of the most thrilling ends to an FA Cup final there's ever been. These were the days when the Cup final really meant something and was the showpiece event of the year for football fans, whoever you happened to support. And here was a pedestrian final apparently decided before half-time by the second Arsenal goal suddenly erupting into barely five minutes of joy, disaster, elation and disbelief, during which emotions were sent soaring and plunging, veering between both sets of supporters to their wildest extremes. All of it captured in the words, voice and authority of Peter Jones.

Jones was the perfect radio sports commentator. He had a combination of talents, from the natural rhythm and pitch of his speech to the ability to find exactly the right words for every occasion, whether it be a dull match with nothing at stake or, like the 1979 FA Cup final, one of those sudden outbursts of extraordinary excitement and drama that only sport can throw up. He possessed authority, erudition and expert knowledge, but at the same time he was clearly a fan who loved the game deeply. The way his voice cracked as Sammy McIlroy's deflected shot squirmed into the net for Manchester United's equaliser demonstrated no pro-United bias, merely a subconscious marvelling at what had just unravelled in front of him. At every moment he was aware of the audience listening at home, aware that he was the conduit between the events on the field and the people hanging onto his words in cars, workplaces and homes across Britain and beyond.

I didn't hear his 1979 FA Cup final commentary live, but Peter Jones provided the first football commentary I can remember. It was a Wednesday in January 1980 and I was nine years old and staying at my grandmother's, tucked into bed with blankets so tight it was like I was deck cargo lashed down ahead of a storm. As a special treat I'd been allowed to listen to that night's football in which Manchester United faced Tottenham Hotspur in an FA Cup third round replay at Old Trafford on Nan's little transistor that she'd placed on a chair next to my bed. I remember nothing about the game itself – I've even just had to look up the score and check exactly when it took place – but in particular I remember a pause in play after United's gap-toothed, mahogany-hard centre forward Joe Jordan had smashed into the Tottenham goalkeeper Milija Aleksic, who had come off by far the worst of the two. There was a long break in proceedings as the players lay on the pitch receiving treatment. Jordan was soon back on his feet but it felt as if Aleksic was down for an age (it would later emerge that he'd lost two teeth and suffered a broken jaw as well as badly damaged knee ligaments).

Peter Jones was the commentator that night and while I don't remember a single specific word he said I do recall vividly the sound of his voice and realising as I lay in the dark that I could actually 'see' what was happening: the prone goalkeeper, the tracksuited physio on one knee tending to him, concerned players standing over them both with their hands on their hips and their breath clouding in the chill air, the bright red shirts of the United players and the white and black of Spurs, even the bare mud of the penalty area where Aleksic lay. With no actual football on which to commentate Peter Jones had conjured the most vivid pictures of the scene in my nine-year-old head that it felt like I was there. It wasn't long before I began to doze, drifting off to sleep to the roar of the crowd mixing with the hiss of medium-wave static and the lilting, lullaby voice of the incomparable Peter Jones.

Something that contributed to making Jones a great commentator was that he became one almost by accident. Born in Swansea in

1930 he read modern languages at Cambridge, earning a football blue before graduating and going into teaching, first at a school local to Cambridge and then at the independent Bradfield school in Berkshire. One day early in 1965 the former Reading footballer Maurice Edelston visited the school and fell into conversation with the language teacher with the soft Welsh accent who seemed to know a great deal about sport. Since hanging up his boots Edelston had been commentating on football and tennis for the BBC, and when Jones expressed his admiration and spoke wistfully of an innate desire to be a sports commentator himself Edelston offered to put a word in for him the next time he was at Broadcasting House.

Within months Jones had left teaching to become an assistant on *Sports Report*, making such progress that when the 1966 World Cup finals came around he was sent to the north-east to report on the group matches at Sunderland and Middlesbrough that featured North Korea, the Soviet Union, Chile and Italy. For one thing, there probably hasn't been a wider range of players' names and pronunciation challenges than there was among those four teams.

Before long Jones had established himself as the voice of the big football occasion, covering every World Cup from 1970 to 1986 and commentating on every FA Cup final between 1971 and 1989. He also covered swimming, Remembrance Sunday and even major royal events: at the wedding of Prince Charles and Diana Spencer Jones broadcast from a pub on Fleet Street alongside Lorraine Chase, which must have been something to hear.

Unfortunately as well as some of football's most spectacular occasions Jones's commentary career also spanned some of football's darkest moments. He had to broadcast live from the Heysel Stadium in 1985 when 39 people were killed at the European Cup final between Juventus and Liverpool after a wall collapsed as people tried to escape a mass stampede on the terraces, and his broadcasts from the unfolding Hillsborough disaster in 1989 remain some of the most compassionate, composed, emotionally restrained and perfectly pitched pieces of journalism in any genre of radio history.

Sports Report dispensed with 'Out of the Blue' that afternoon, a rare and possibly unique incidence of the jaunty theme not opening the programme. Instead, after a brief outline of the dreadful events in Yorkshire, the studio handed to Peter Jones at a Hillsborough stadium now almost completely empty of people, where he remained in the commentary position from which he'd expected to summarise a thrilling cup semi-final. Instead, having provided regular bulletins of unfolding horror for the previous two hours, he found himself having to reflect on arguably the most appalling tragedy of modern times. From the gantry he'd watched the supporters stream into the ground in the sunshine of a warm April day, the buzz of a big game making the air crackle as the songs of both sets of supporters washed back and forth across the empty pitch. He'd watched the players come out to warm up and seen the lower terrace of the Leppings Lane end become more densely packed as kick-off approached. He'd seen crowded terracing before but, as the game began, from his elevated position in the commentary box, this looked different. In the two central pens behind the goal there were just faces, no shoulders, no arms, no hands clapping above heads – in fact hardly any movement at all, just tightly packed faces.

He'd seen the first supporters managing to climb out of the overcrowded pens onto the pitchside and people hauled up to the top tier by fans desperately trying to help relieve the crush. He'd seen the match suspended after barely six minutes when a policeman ran onto the pitch to inform the referee something dreadful was happening, and then he'd watched scores of people die in front of him, people who'd left home that morning to watch a football match but would never return.

All afternoon he'd given out updates of the rising number of confirmed dead and in one report described how an ashen-faced young Liverpool supporter, his face wet with tears, had climbed the steps to the commentary gantry to ask if he could use the telephone to call his mother. 'Of course he can,' said Jones, so lost in the awful tragedy encapsulated by that moment he slipped into the present tense, as if it was happening right then. 'Of course he can,' he repeated.

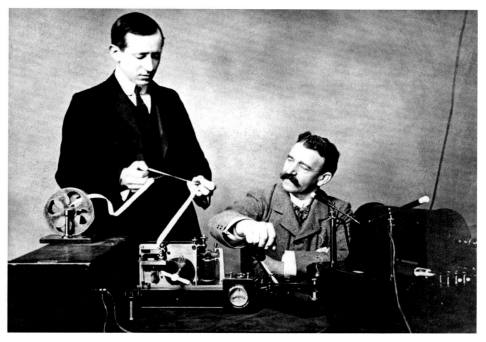

'It finishes with, "…and a big hello to anyone who knows me".' Guglielmo Marconi (*left*) and his assistant George Kemp with an early wireless message. Kemp is ready with a big lever because it wouldn't be proper science without a big lever. © *Time Life Pictures/Getty Images*

Young people today have lost the art of conversation, just sitting around with their earphones in the whole time. An 'Electrophone salon' in London's Gerrard Street, circa 1903. Note the evocative mural of Cupid being electrocuted. © *Print Collector/Getty Images*

The understated tribute to Guglielmo Marconi, tucked away outside the Haven Hotel, Sandbanks, Dorset. Calling him 'the inventor of a wireless telegraphy system' is a bit like calling Tim Berners-Lee 'the inventor of a computerised information network'.

Chelmsford, 1920: Dame Nellie Melba makes the first scheduled British radio broadcast with the help of an old hat stand and some bits of cigar box. Also a tiny handbag.
© *Universal History Archive/Getty Images*

In her garden at Oxted, Surrey, cellist Beatrice Harrison accompanies a nightingale for the nation's listening pleasure. By coincidence the bird was a direct ancestor of Radio 1's longest serving presenter, Annie Nightingale.
© *Topical Press Agency/Getty Images*

Lord Reith, the frankly terrifying first Director-General of the BBC. The man on the right is probably about to regret laughing at his shoes. © *Fox Photos/Getty Images*

Pioneering documentary-maker Olive Shapley (*left*) tries to remember where she parked the seven-ton outside broadcast truck in Craghead, Co. Durham, 1939. © *BBC Photo Library*

Britain's first sports commentator Teddy Wakelam at Wimbledon in 1935, possibly on the very day he set light to his trousers mid-broadcast. © *BBC Photo Library*

The magnificent Shelia Borrett captured at the exact moment she heard someone say, 'it's simply not a job for a woman'. © *BBC Photo Library*

Clapham and Dwyer, the double act who in 1935 scandalised the listening public with a filthy joke, attempt to distract from their wanton corruption of the nation's morals by pointing at a picture of a cow.

The father of independent radio, Captain Leonard Plugge, pictured in 1936. He eventually gave up trying to tell people it was pronounced 'Plooje' and is pictured here in a tuxedooje with a flower in his buttonhooje.
© *Fox Photos/Getty Images*

The headquarters of Britain's smallest radio station Two Lochs Radio in Gairloch, Wester Ross. It's so small inside that you can't even say 'Two Lochs Radio', you have to say '2LR'.

Charlotte Green, custodian of the national sporting gazetteer that is the reading of the football results. © *BBC Photo Library*

5 Live's Dotun Adebayo, the busiest man in radio, somehow right on top of his game at twenty past four in the morning.

Corrie Corfield, the shipping forecast, fair, good.

Hilversum's Media Park needs all those signs because you'd never notice that unassuming building otherwise.

A postcard received from the radio.

End of transmission: what's left of the Rampisham Down transmitting station in west Dorset from where voices were once catapulted right around the world.

His final summary would take little more than a minute and consist of no more than three hundred words, but it's a piece of broadcasting still able to evoke the enormity of the Hillsborough disaster even today. Having digested the day's events as much as he possibly could, when the *Sports Report* studio handed back to him once again Jones began by describing the irony that the sun was shining on Hillsborough and the green Yorkshire hills beyond, and wondered aloud how dozens of people could possibly have died in the now peaceful, sun-drenched stadium where he sat. He spoke movingly of the Heysel tragedy, barely four years earlier. He told of how the Hillsborough gymnasium was being used as a mortuary, and watched the stadium stewards walking dazedly among the twisted crush barriers of the Leppings Lane pens with boxes and bags, collecting up the discarded personal belongings that littered the concrete steps, the hats, scarves and rosettes, the favours of Liverpool, the club they'd come to support.

He concluded by saying, simply, 'And the sun shines now.'

Six weeks later Jones was at Anfield commentating on the match between Liverpool and Arsenal that would decide the destination of the league title. Arsenal needed to win by two clear goals to snatch the title from Liverpool but, for once, in the circumstances, everyone in the country who wasn't an Arsenal fan wanted Liverpool to prevail.

As the game entered its final moments Arsenal led 1–0 and the title was seconds away from being Liverpool's. Then the Arsenal midfielder Michael Thomas picked up the ball and began dancing through the Liverpool defence on the edge of the penalty area. Famously on television the commentator Brian Moore hooted, 'It's up for grabs now!' as Thomas stumbled towards goal.

On the radio Peter Jones said, '. . . and Thomas goes inside Nicol, and Thomas is there . . . and Thomas has scored for Arsenal!'

The voice reached its crescendo and rasped slightly in his throat as it had done at Wembley almost ten years to the day earlier, but all the joy had gone. The delighted disbelief of the spectacle that had underpinned his commentary on the climax to that 1979 final simply wasn't there any more. His voice rose in line with what he was seeing, arguably the most exciting climax to an English league season in history, but when it fell again as he described the aftermath of the goal there was a tangible sadness in the cadence.

Peter Jones was dead within a year. On 31 March 1990 he was commentating on the build-up to the Boat Race from the BBC launch the *Arethusa*. As he ran through the details of the two crews lined up on the water in front of him Jones suddenly seemed to falter, his words deserting him, and after brief silence his colleague Dan Topolski took over. Jones, sitting in his commentary position with the microphone at his lips, had suffered a massive heart attack. He died a day and a half later in hospital. He was 60 years old.

In the last five years of his life Peter Jones had been present for two of the worst disasters in modern British history. Ninety-six people died at Hillsborough and 39 at Heysel: Jones had effectively witnessed the equivalent of two serious train crashes, or a pair of major terrorist bombings. Not only that, he'd had to watch every moment, broadcasting descriptions of the events as they unfolded and then summarising them afterwards. In both cases he'd only been there to commentate on a football match, to do a job he loved for a sport he loved for the benefit of an audience who loved the game as much as he did.

Whether Hillsborough had broken his spirit is impossible to say, but when Peter Jones died a particular form of commentary died with him, one at which he was the master-craftsman. He had come from a world outside sport and outside the media. He was widely read, packing volumes of Dickens and Hazlitt on flights to World Cups. He knew the importance of sport but also knew there was more to life. He had no desire to project his personality on his listeners, to be sensational or controversial; he just sought to describe events as they happened for the benefit of listeners at home. And, my goodness, he was extraordinarily good at that.

15

'Make it a Conversation': Spending the Night with Dotun Adebayo

'If you make through to half-two I'll be impressed.'

I was standing at a drinks station in New Broadcasting House with Dotun Adebayo. It was midnight, Sunday night turning into Monday morning, and he was doubting my ability to make it through the night without nodding off. Outside in the darkness central London was about as quiet as it would ever be, and so was the building in which we were both opening cupboards in search of teaspoons. The place was absolutely deserted. With the darkness outside it felt a little like we were a skeleton crew on the bridge of a huge spaceship transporting cryogenically-sleeping colonists through the furthest spiral arm of the galaxy. And someone had left the teaspoons back on Earth.

We met a few weeks after Dotun Adebayo had been acclaimed as Best Speech Presenter at the Audio Production Awards. He was genuinely surprised and authentically dumbstruck when he reached the stage, improvising an emotional speech during which he broke down in tears. If ever an entire room full of people metaphorically reached out and gave someone a hug, it was at that moment. 'It's funny,' he said, reversing backside first out of the cupboard where he'd been foraging for spoons, 'it's only a short time since that award but already I can tell that it's made me much more confident in myself as a broadcaster. I'd never won anything like that before.'

Dotun is the busiest man in radio, working seven nights a week, every week, presenting the overnight show on BBC Radio London from 1 a.m. to 4 a.m. on four nights, then *Up All Night* on 5 Live between 1 a.m. and 5 a.m. on the other three. For the national show he spends two nights broadcasting from Media City in Salford, renting a tiny, cheap flat in a rough part of Manchester, where he grabs a few hours' sleep during the day, and Sunday nights in London because on Sunday evenings he also presents a two-hour magazine show aimed at the capital's black community. That's 26 hours of live radio every week, on top of being a newspaper columnist and running his own successful publishing company X Press (the imprint's first book, 1992's *Yardie* by Victor Headley, is being turned into a film starring Idris Elba, he tells me). Oh, and he's a husband and father too, with one daughter already away at university and the other about to start.

He's been presenting *Up All Night* since the early part of the millennium, taking the weekend slot while the show's founding presenter Rhod Sharp looks after the weekdays. He took over the Radio London show, and began working seven nights a week, in August 2017.

'A colleague of mine at Radio London was away for a while and they asked me to sit in for a month or so,' he told me, turning on a tap and rinsing a spoon he'd found, stirring our cups of tea before setting off towards the 5 Live studio. 'He was quite ill though, and in the end he didn't want to come back so they asked if I'd like to do it permanently. I jumped at the chance. That programme has had my name written all over it for years and finally when the chance comes I'm already doing three nights a week. It worked out perfectly, though – the four London nights and the three 5 Live nights fell into place like a jigsaw.'

Dotun is married to the singer Carroll Thompson, and surely, I wondered, they must hardly see each other.

'Well, being a musician my wife has a different lifestyle too – neither of us do the nine-to-five thing. So we have to catch our moments. We do find time to go out but you really have to make time for that, and it does sharpen your obligations. I'm not that bothered about travelling or partying these days. I did plenty of that when I was younger.'

We were walking through a succession of deserted open-plan offices, rows of desks that gave the impression everyone had just stepped out for a moment – notebooks were open, post-it notes stuck on screens, the odd paper cup and coffee mug lying around. It felt almost like a room full of ghosts, but then I realised that it was Dotun and me who were the ghosts, the creatures of the night; the people who spend the daylight hours at these desks would never know we were there. When we reached the studio – the same one where I'd watched Charlotte Green read the football results – Dotun put down his mug of herbal tea and the tray of chips he'd squired from the canteen and we settled ourselves in for the night.

Radio comes into its own at night, somehow. There's a real sense that you're the only one listening, that it's just you and the presenter awake, keeping a night watch for the world. Traditionally, small-hours radio in Britain has been speech-based and usually there's one presenter: no double acts so less banter, an authentic one-to-one conversation with the listener. Overnight radio is also less tied to rigid slots and timings, allowing more in the way of spontaneity – radio expanding into the night for its audience of insomniacs, night workers and misfits.

'There are a lot of shift workers, lorry drivers, taxi drivers,' replied Dotun when I asked about the make-up of his audience. 'New mothers being kept up by their babies, and we have a lot of overseas listeners in different time zones too. We also have a lot of lonely people listening who need someone to talk to even if it's just the radio. They're party to our conversation and not alone any more.'

For somewhere crammed with people the city can be a lonely, isolating place. Radio is a useful haven for the socially withdrawn at any time of day but particularly at night. Darkness encourages introversion, and if there's a voice coming out of a box on the table – a friendly voice from a warm personality, engaging and entertaining in a manner appropriate to the hour and the audience – then radio, and public service radio in particular, is doing its job properly.

Famously there were occasions in the late eighties and early nineties when Peter Cook, practically a recluse at that stage of his life, would call Clive Bull's late night show on LBC pretending to be 'Sven', a Norwegian who wanted to talk about his lost love Yuta, and fish. There are also extreme tales like that of the night Adrian Love, who presented overnight shows on LBC and Capital Radio during the seventies, left the studio with an LP playing to go and find a young woman who'd called in to the show saying she'd taken an overdose – he found her on the street in her nightdress and was able to escort her to hospital where she was able to receive the treatment she needed. The Radio City overnight presenter Peter Price did almost exactly the same thing in 2004, heading into the streets of Liverpool in the middle of the night to track down a suicidal 13-year-old boy who'd called in to his programme. That's an astonishing level of intimacy between broadcaster and audience, one that wouldn't (couldn't) be replicated during the day.

'I think a lot of people regard us as confessors,' said Love after his mercy dash. 'With radio they know that whatever they talk about they can, if they wish, remain completely anonymous. Nothing else, not even the parish priest, is as private.'

Now, a broadcaster could never take the place of a trained psychiatric professional, but as a companion through the loneliest hours the late-night radio presenter provides an invaluable service. Unfortunately the relatively low audiences mean that many stations have cut their overnight programmes, but BBC local stations across Britain take *Up All Night*, meaning Dotun and Rhod Sharp pull in a considerable audience for the antisocial hours they broadcast.

'On any one night there's probably a minimum of 300,000 people listening, which is a lot for the middle of the night,' Dotun said, between chips. 'This show probably has the biggest share of the market: something like one in five people listening to the radio at that time of night are listening to me, and that's a privilege, man, that's a real privilege.'

It's not the largest audience in the radio firmament but it's probably the most loyal, a community unto itself in many respects. Just as we'd ghosted through the deserted open plans on the way to the studio, night-time radio is almost a parallel service in a parallel world. *Up All Night* is a different beast altogether to Dotun's Radio London Sunday show, where his working night had begun four hours earlier in a different part of the New Broadcasting House complex. The BBC's London station has undergone a number of transformations over the years, not always to its benefit, but its dedicated programming aimed at London's black community is one thing that has remained constant.

Dotun had breezed into the building barely ten minutes before he went on air, then conducted two hours of radio that I realised as I sat behind him in the studio was a masterclass of its kind. He's completely alone in the studio presenting his 5 Live show – on a Sunday his producer is still in Salford, appearing occasionally as a disembodied voice down the line – but for that evening's Radio London programme he was joined in the studio by a panel of six people, with two members of the production staff in the control room on the other side of the glass.

The programme's theme was the abolition of the notorious 'Form 696' after a 12-year battle with the powers that be. Since 2005 venues and promoters putting on events featuring DJs and MCs had to submit the form to the Metropolitan Police, ostensibly as part of a risk-assessment process but identified by London's rap and grime communities as specific targeting of their events because they're attended by a mainly black audience.

Arranged round one side of the console to discuss Form 696 and the grime scene in general was a diverse group ranging from academia to grime artists themselves. Other than Mykaell Riley, the University of Westminster's head of music production and a former member of the Birmingham reggae band Steel Pulse, none of Dotun's guests were remotely used to being in a radio studio, yet within minutes he'd drawn them into a discussion in which the microphones in front of them were practically forgotten. He was careful to ensure everyone

175

felt included, coaxing them out of any nervous reticence and giving each a fair share of microphone time, remaining conscious all the while of the fact that grime artist Stormin MC was in the advanced stages of skin cancer – he died in hospice care barely eight weeks later – so required extra-sensitive handling.

Dotun has an in-built natural radio barometer, by which I mean he knows instinctively when to play the get-with-it-Grandad old timer bewildered by the young people and all their noisy pop music, and when to be the leading light of London's black cultural community that he unquestionably is. Within the space of a few minutes at the start of the programme he'd paid an emotional and spontaneous tribute to the reggae artist Michael Prophet who'd died the previous day and told the story of a conversation about calypso music he'd once had with the Queen. Shortly after that he was out of his chair dancing to a grime track, arms flying, the lead from his headphones flapping and snapping like a halyard in a force nine.

All the time, even while watching him effortlessly conduct six people from a range of backgrounds, most of them strangers to each other, one of them terminally ill, it was clear that Dotun always had the listening audience at the front of his mind. When he took a call from 'Tony' who informed the panel that their work was not music for 'normal people' and, anyway 'you can't make real music on a computer', Dotun listened, engaged, kept him on the line and played a trail during which he encouraged his studio guests to respond to and engage with Tony despite their understandably levitating hackles. 'Make sure you address him by name,' he advised them. 'Make it a conversation.'

Sure enough, what had begun with a rant that could easily have been answered with a rant turned into a measured and enlightened discussion. From unpromising beginnings – many presenters would have rolled their eyes and cut Tony off pronto – Dotun had created a terrific piece of radio, one conducted with the audience uppermost in his mind. It was characteristic of a broadcasting style that doesn't second guess the audience, doesn't assume knowledge but doesn't patronise or talk down either. Pitching that right is some trick if you can pull it off. You really need to know your audience.

'It's a black-oriented show,' said Dotun afterwards as we headed from the Radio London studios across to 5 Live, 'but I think of the white guy listening who has a mixed-race daughter and wants to find out more about black culture.'

It's these levels of empathy and warmth that combine with a fierce intelligence to help make Dotun Adebayo a terrific broadcaster. He's a man with an innate respect for his audience and a natural feel for the medium, and someone who has a wealth of life experience to draw on. It's as if he was constructed specifically for radio.

'My earliest radio memory is of my grandmother's house in Nigeria, a big place in a compound where my brothers and I all lived before we came to join my parents in England,' he told me, finishing his chips as we waited for 1 a.m. to tick round.

'People had speakers on the outside of their houses in those days, tuned either to the Nigerian national broadcaster NBC, where my father used to work as a presenter, or the BBC World Service. I'd have been about five years old so this is around 1964 or 1965, the early days of Nigerian independence and a time when you needed to know quickly if anything dramatic was happening. I distinctly remember hearing on NBC that there'd been a coup. It was definitely some important political news involving the military.

'After I came to Britain I had a transistor radio from the age of about 11 that I definitely didn't buy because I wouldn't have had the money. Maybe a friend had nicked it, that was the kind of background I had, and I probably paid him ten pence for it, which was a whole week's pocket money in those days. I used to listen to Radio Luxembourg on that radio, under my pillow late into the night.

'A couple of years later Capital Radio had a phone-in show where you could call in and speak to whichever semi-celebrity was on that week. Around that time I'd made up my mind I was going to be a pop star and, on one particular night, the studio guest was Dave Dee from Dave Dee, Dozy, Beaky, Mick and Tich. I didn't really know or care much about him or his music but I went to the old payphone we had in the hallway of the shared house we lived in and was shoving two-pence pieces into the slot while telling Dave Dee that I wanted to be a pop star too.'

Whatever advice Dave Dee dished out to the young pop hopeful on the end of the line, it was either duff or just fell on stony ground, because Dotun's route to radio came via a different career altogether.

'I was an actor before I was anything else. Bizarrely, when I was eight years old I was in a Hammer Horror film called *The Oblong Box* in which my role was to be run over by Vincent Price on a horse. With that under my belt I thought I was well on my way to being an actor. The only problem was that I was rubbish at acting. I didn't realise this until I was about 20 years old at the National Youth Theatre with guys like Timothy Spall, who took it much more seriously and had a level of commitment that I didn't have.

'I got into radio initially in quite an unusual way. I don't know if I thought it would help me succeed as an actor, or what my motivation was, but while I was still at school I presented myself at the World Service, asking for a job. My father had come to Britain in 1964 to be a scientist, a period when people didn't respect what you might have achieved elsewhere. It was a race thing to a certain extent, but I think it was more a colonial thing. "You were a broadcaster in Nigeria, eh? So what?" He'd actually been a big cheese there, they called him "the angry young man", and there was even a cartoon of him published in one of the Nigerian national papers that had him sitting behind a big old fashioned microphone being asked "What are you so angry about?" and him replying, "Anything! Everything!" like Marlon Brando in *The Wild One*.

'He struggled to get work at the World Service and did bits and pieces here and there between his studies. I was about 12, still set on being an actor, when my father suggested I go down and see the editor of Network Africa. So I took the [old] 171 bus from Tottenham to the Aldwych, walked to Bush House, found the Network Africa office and parked myself in front of the guy in charge. I said I was Lai Adebayo's son and was there because

I wanted to work for him. At 12 years old! He looked me up and down and must have been thinking, "You're having a laugh, son." At the same time there were telexes coming through into the office, a lot of it in French. I said "Oh, I could translate that for you." Which I couldn't, really, but he saw how keen I was and said, "OK, come in in the school holidays." I used to spend most of my holidays at the World Service, just helping out.'

We'd reached the time for the show to begin. The first half hour or so was an item pre-recorded as live before we'd started hunting teaspoons, a discussion with a correspondent in Beirut about the murder of a British diplomat there. It was a wide-ranging conversation, one that covered the political aspects and the tragedy's potential effects on diplomatic relations, and also the impact on a more personal level: the correspondent had met the victim not long before the end of her life. This is where the overnight feel of the show comes into its own: the expansive format allows an immersion in a topic, a depth and infusion with nuance that the pacier nature of daytime radio won't permit.

'The old format of *Up All Night* used to be a little bit different,' said Dotun. 'In the first hour of the show there would be maybe a dozen different items, but I've started giving the first item half an hour before going on to other topics. Usually, instead of talking to just one correspondent as we are tonight we'd have two people coming at it from different perspectives. Not necessarily to have a ding-dong, just to give a rounded impression of the subject we're talking about. There was a remarkable night in Riyadh recently when 12 Saudi princes were arrested, then the Lebanese Prime Minister decided to resign in Riyadh, and after that they apparently shot down a missile fired from Yemen – over Riyadh. Initially we were speaking to two academics, one after another. The first was a Saudi expert, the second specialised in Lebanon, and just as I was speaking to the first guy the producer said, "Let's try putting them together." I had to ask them on air if they were happy to open it up to a discussion and luckily they were both up for it. With them being academics they were very polite in their disagreements – all, "I agree with most of what you're saying, but . . ." – that kind of thing, but with the way the stories were developing while we were on the air it made great radio.

'I think good radio comes about through a harmonisation of voices, not because those voices are necessarily agreeing with each other, rather they're harmonising around a particular topic. Something nobody else had considered before that night was Lebanon being the new front line of a proxy war between Iran and Saudi Arabia. Nobody had put out that line of thinking until one of these guys said it and the other agreed with him. That way we got something from our coverage that had never been mentioned in the debate before.'

As the recorded interview played out quietly over the speakers Dotun picked up the story from his youthful World Service days.

'After that I had no real contact with radio until I went to study in Stockholm. I started writing for a music magazine there and was introduced to a friend of the paste-up artist, the guy who laid out the pages, who worked at the Swedish equivalent of Radio 4. P3, I think it was called. He'd seen me writing about reggae in this music paper and asked me if I would do a slot called "the week's reggae" on one of the music shows, in Swedish, in which I recommended a reggae track every week. I started coming up with programme ideas, made a documentary on the death of Bob Marley in 1981, and did a few other bits here and there. I came back to Britain in 1983 and went to university in Colchester where I did some stuff at University Radio Essex. Essex was ahead of the game in those days, one of the first universities to have its own student radio station, but now I go to student radio award ceremonies and there's no sign of Essex anywhere, so I don't know what's happened there, it's a real shame. I did a couple of shows on URE, one I called something like *The Cotton Candy and Hamburger Show*. I don't know what the hell that was about.

'Sometimes it got heavy, as it happens, because I had these racists who kept calling up. It was partly a phone-in show, and these same two or three people kept calling in giving it "you black bastard" and all that. I was pretty shocked but thought, "OK, I can handle this." There were some heavy duty Irish republicans up there at URE in those days and this bloke with a thick Northern Irish accent insisted I put him on and he said something like, "To the guys making these racist comments, we know exactly who you are and we are coming for you." It did make university a bit uncomfortable for a while, walking around the campus thinking about how these racists were somewhere among all the people milling around, walking past me right now, probably laughing their heads off. I ended up being elected president of the student union, though. I was there at the same time as John Bercow, the Speaker of the House of Commons, and I won with the biggest majority Essex had ever seen. I resigned at the end of Freshers' Week, but that's a story for another time.'

As the night progressed I realised just what a global remit *Up All Night* has. After the Beirut item there were reactions to a general election in Chile, a live chat with a correspondent in Buenos Aires about a missing Argentinean submarine, a conversation about the Harvey Weinstein scandal with a writer in New York, and analysis from Washington of a terrorist plot in Moscow that was foiled with American help – all the while handing over to Australia for updates on the Test Match cricket. I shouldn't really have been surprised: given the time-zone situation a global perspective on the news is perhaps inevitable: in great swathes of the world we were well within regular working hours for the correspondents.

'After university I worked on newspapers like *The Voice* and started my publishing company X Press, which kept me busy. As a result I'd often be asked to comment on events of the day on GLR (as BBC Radio London was called back then). Some time around 1993 Gloria Abramov, the station editor, asked me in for a chat. She'd seen me come in to do an interview on one of the shows and it was a case of being in the right place at the right time: the person doing their black interest programme had gone AWOL and they needed someone quite quickly. She asked me if I'd do it and I said, "Yeah, sure, I'll do it." I got 50 quid a programme and stuck at it even though by then X Press was really starting to take off. No matter how busy I was, radio had me hooked.

'In 1998 I was about to become a father for the first time and the *Guardian* asked me to write a piece on the experience. Somebody at 5 Live, a station I'd never listened to in my life, saw what I'd written and asked me to come in as a guest to talk about it on one of their shows, and before I knew it I was coming in all the time talking about different things. My publishing background got me a gig as a regular guest on Simon Mayo's programme talking about books, for example. I was doing so much talking-head stuff the ex-boyfriend of a girl I was going out with at the time was pissed off because every time he switched on the radio "that bloody Dotun" was on!

'Then around the turn of the millennium I had a call from Bob Shennan, 5 Live's station controller, asking me if I'd meet him for lunch to talk about some programme ideas. By then I'd started doing the odd shift sitting in on their obituary programme *Brief Lives*, and he offered it to me permanently. I said I appreciated it but what I really wanted to do was *Up All Night*, which must have taken him aback because most people regard this as the graveyard shift. Still, a couple of days later he called me and gave me my first overnight shift, and I've more or less been here ever since.'

At around two in the morning – approaching the time by which Dotun had predicted I'd have been slumped and dribbling (but I was still giving at least the impression of being awake) – there was an explosion of noise in the corridor outside. One of his regular slots is Generation Gap, when two women in their early twenties come in once a month to discuss an issue relevant to their generation. They bustled into the studio with a fizz and vigour that would last throughout the segment and seemed to someone whose eyelids were becoming weighty frankly out of place for two o'clock on a Monday morning. Their discussion topic for the night was love and relationships, a 90-minute conversation that covered topics as varied as falling in love, jealousy, coming out and pornography. Millennials get an incredibly hard time from older generations but here were two intelligent, informed and articulate women who could talk about Snapchat with the same nuance, depth and engagement they employed when discussing sexual consent. It was wildly impressive stuff. They fielded calls from listeners and opened up about their own romantic lives and encounters with astonishing frankness, honesty and, at one point, even tears. It was an hour and a half of tremendous radio, again exemplifying the space and depth permitted by the relaxed overnight format: during the day they'd have probably got 20 minutes at most, and even that would have been interrupted by the travel news. At the end of the item there was a flurry of hugs and they bustled off back into the night, still talking, still full of energy.

'They're brilliant, aren't they?' said Dotun. 'That was great radio. I can pretty much just give them the topic and let them go, which is great for me. I'd say that by profession and instinct I'm a newspaper journalist, but working in radio has taken that innate news journalist in me somewhere quite different, and that's a great example. Radio is a different skill to written journalism, and I suppose some of my acting background comes into it because I think presenting is an acting gig, basically. I don't mean that it's not real, but a radio presenter is a messenger. With news broadcasting you've got to get the message right and you've got to get it across clearly, and in items like that it's steering the speakers as and when they need it, making it as easy as possible for them to do their thing.'

Even before their arrival I'd been surprised at how busy and vibrant the show had been. I don't know what I'd expected – a studio darkened but for a pool of light from a desk lamp, perhaps, where Dotun murmured into the microphone in response to a succession of lonely callers? Instead this was a pacy, varied show pulling in correspondents from all over the world and featuring lively debate across a raft of topics. Four hours is a long time in radio – you won't find many single-presenter shows of that length in the schedules. That, combined with his other shows, makes Dotun's a feat of broadcasting stamina that few could match, yet he makes the whole thing seem utterly effortless.

By 3.30 a.m. the rush of adrenaline provided by the two women had evaporated, at least for me. I was starting to fade, but Dotun was in his element, chatting to a reviewer about the week's best podcasts. During the four o'clock news he turned to me and informed me he was so impressed by my staying power he intended to treat me to breakfast in the canteen afterwards. If I was still awake at that stage. The last hour could have dragged but it absolutely didn't as Dotun rattled through a controversy regarding the new iPhone's face-recognition software and a preview of the forthcoming BBC Sports Personality of the Year event. At five o'clock Dotun wrapped up the show, thanked his producer down the line in Salford and handed over to *Wake Up to Money*. His night's work – all nine hours of it, including six hours of live broadcasting – was done.

We walked back through the empty offices to the canteen where early shift journalists and presenters were starting to trickle in, heading blearily but determinedly for the coffee machines like the walking dead. Having passed along the counter, with eggs, beans, sausages and bacon piled onto plates, we sat in a booth and I asked Dotun what he thought makes good radio.

'In terms of how you present there's a certain cadence in the voice you need to master. It's so important to be clear in the way you speak and the way you deliver the message, and you do have to act a little bit, pitching your rhythm, pace and fluidity so it's just right. At a newspaper you can sit down and pontificate into the keyboard and the reader has as many times as they need to read it and understand it. On radio you only have one shot so you have to get it right first time.

'The reason I go up and down to Manchester every weekend is because *Up All Night* is the perfect programme for the type of broadcasting I do. I'm aware of my limitations and have to play to my strengths. I wouldn't be as good doing, say, a breakfast show, but this kind of global mixed-news magazine programme is right up my street. I'll never get an opportunity like this anywhere else, it's only here. If I hadn't agreed to go up to Manchester for 5 Live I'd have regretted it for the rest of my life.

'I'm not blowing my own trumpet here, but the reason I got that award is not because I'm better than anyone else, it's down to the fact I do programmes that don't show my broadcasting weaknesses. If they threw the *Today* programme at me it would be a disaster, I'd be completely exposed, my weaknesses would be there for all to hear, but I can do *Up All Night* almost with my eyes closed. Not that I do it off the cuff, not by any stretch of the imagination, I put the work in; I really enjoy the research and the preparation. I absolutely put my heart and soul into this programme.'

One of the most impressive aspects of watching Dotun in action was the total absence of ego. He was authoritative without being overbearing, and when in conversation kept his interjections to a minimum, letting his guests and correspondents speak. The programmes I'd witnessed over the previous night were entirely different in tone and content yet both were stamped indelibly with his personality. If anyone deserves to have an ego it's Dotun Adebayo, but there's really no trace of one. At one point over breakfast I explained the process of writing and publishing a book and he nodded with genuine interest. It was only on the train home that I remembered with a wince how he'd launched and run one of country's most successful independent publishing houses. It's that kind of self-effacing warmth that's helped to create such a good broadcaster, as well as the constant presence in his mind of the people listening at home.

'I never lose sight of what an honour it is to be part of that community, to have that relationship between presenter and listener. On the Radio London overnight, in particular, you know everybody: 80 per cent of the callers are regulars. There's even one old guy from my school who calls up and we start singing the old school song. I regard it as an absolute privilege to be allowed into these people's homes. Television is more a decision of choice. It might not seem that way, but switch on the television and most of the time you don't know what's going to be on – but switch on the radio and it's more or less the same schedule and same voices every single day. There's breakfast, a morning show, an afternoon show, drivetime, the evening slots, the same presenters day after day, whereas with TV you're seeing different faces and hearing different voices every time you switch on. Radio is a very conservative medium, in a sense, and that's something people don't want to change. Radio provides the rhythm of people's daily lives, and they like to have the same voices there when they're supposed to be [there]. It's a very subtle science, radio, based purely on the voice.

'You have to understand your audience. You've been invited into their homes so you have to ask yourself how you should conduct yourself. You behave in a different way in your friend's home from, say, when you're in your elderly neighbour's home. You have to pitch yourself right, hitting the median of whoever your audience might be. If it's an older audience then it's no use using vulgarities, or talking about sex or death because, well, older generations spend enough time thinking about that last one already. It's not so much about who they are as what *engages* them. Like in the London show earlier, if you engage a young audience by having a debate about grime, they're on it. A phone-in tells you exactly who the audience is, and from the phone-in you see who they respond to, and that's great radio.'

For all the rewards, though, don't the hours ever get him down? Working seven days a week would leave most people shattered and dispirited, let alone seven nights.

'My sleep pattern doesn't shift from overnights to daytime,' he explained. 'When I was just doing the three nights a week I was still staying up all night on the other four, home by myself, everyone asleep, trying to do a bit of work for my publishing company here, trying to read a book there, essentially alone and not remotely sleepy, just sitting downstairs twiddling my thumbs. That was much harder for me than working seven nights. Now I'm doing something I enjoy instead: presenting radio programmes. When I come out of the studio at five for this show and at four on the Radio London show I'll go straight to the gym for a good two hours or more. I'm fitter now than I have been for a long time. Even when I'm in Manchester my first train back after finishing my shift at five isn't until eight o'clock, so after the show I jump in a taxi from Media City to the gym nearest the railway station. It all works perfectly.'

It's not all cheery debates with millennials and chats about podcasts, however. In the months before we met Dotun had been at the sharp end of some of the nation's most heartbreaking stories.

'I was on air for the Manchester Arena bombing,' he told me, looking down at his empty plate and mashing his napkin in his hand at the memory. 'I was in the Manchester studio getting things ready at about ten o'clock when my producer said there were reports of an explosion. The news kept filtering in and it was just awful. That was a hard one. No news story I can think of was as painful as that one. Little kids going to watch a concert. That's just the worst kind of horror, seeing footage of the young girls screaming and trying to get away, absolutely terrified. I was on for the London Bridge attacks too, and I was over in west London talking to people the evening after the Grenfell Tower fire. It was an eerie feeling, not least because the top floors were still glowing. When that kind of thing happens you just have to scrap everything else and it becomes a difficult show, partly because you have to work quite hard at keeping a consistent, balanced and paced delivery all the way through, no matter what's happening or how the incident is developing.'

It was nearly 6 a.m. Despite the fact he'd travelled down from Manchester the previous day, worked for nine hours, would have been in the gym if he hadn't bought me breakfast, and hadn't seen his family since before the weekend, Dotun insisted on giving me a lift to the station. We drove through the still-dark streets that were just starting to get busy with commuters even at that early hour.

'At some point a younger person will rightfully take over from me and I'll be pensioned off,' he said as he pulled in to let me out. 'And you know what? When that day comes I will look back and think I absolutely gave it my best shot. No regrets. Not a single one.'

16

Last Heard of: the Lost World of the Radio SOS

'Here is an appeal from the Middlesex Hospital. If anyone is listening in from Ampthill, Bedfordshire, will he please inform Mrs Carr of Thinnings, Flitwick, that her son is very ill at Middlesex Hospital. Will she please come at once.'

This message, broadcast just before 9 p.m. on 29 April 1923, prior to an address by the Rev P.T.R. Kirk of the Industrial Christian Fellowship, wasn't quite the first SOS message broadcast on British radio but it was the earliest in a format that would become familiar to radio listeners for the rest of the twentieth century.

The broadcast SOS hasn't made it into the twenty-first, however, and while it doesn't seem as if a decision was ever taken by the BBC to stop putting them out, the advent and proliferation of social media and mobile phones is the likeliest explanation for their absence from the airwaves. So subtle and unheralded was their demise that nobody is quite sure when the last one actually went out, with most radio people seeming to agree there hasn't been one this side of the millennium at least. But for almost 80 years such appeals were a regular interruption to the schedules, tiny slices of raw human drama and tragedy.

In the earliest days the messages were not just confined to relatives of the gravely ill: there were appeals for missing persons, and police requests for potential witnesses to accidents and even murders. So regular were they, in fact, that between the 1930s and 1950s there were often more than a thousand SOS alerts broadcast by the BBC every year; almost three a day. In 1929 881 SOS messages were broadcast, of which 317 were recorded by the BBC as 'successful'. For all the messages that were put out there were hundreds more that were submitted but never made it to air. There was the woman who requested an appeal for a lost teddy bear named Benny, for example, meandering through London somewhere on a number 9 bus. There was the set of false teeth believed lost in a swimming pool, the woman whose dancing partner had broken his leg and wanted an SOS broadcast for a replacement, and the frankly unsavoury sounding man offering £5 to anyone who would 'bring my wife home for a good hiding'.

The messages became more frequent during the summer months when people were away on holiday. The increase in leisure motoring that had begun in the late twenties led to a much more peripatetic summer population, hence the surge in pleas for people 'believed to be currently driving a green Austin 8 in Cornwall' to call home urgently.

The appeals were never repeated: once they'd been read out they were gone for ever and listeners would never learn whether Arthur Robinson, last heard of two years ago in Chester-le-Street, made it in time to visit his father in Leeds infirmary where he was dangerously ill. The fact that in the early decades of radio more than half the broadcasts were regarded by the BBC as having been successful is a decent return, and demonstrated how far radio had disseminated into British society. If the person concerned didn't hear the message there was a strong chance that someone who knew their whereabouts did.

The brief pleas for a particular person to contact their families produced some poignant radio. Contained within those straightforward, no-nonsense appeals were heartbreaking glimpses into personal tragedy. The thought that someone was driving around the country, roof down, the wind in their hair and a *Shell Guide* on the passenger seat next to them, blissfully unaware of frantic attempts to contact them couldn't fail to tug at the heartstrings – a sobering reminder of both our mortality and that of the people closest to us. You knew the appeals were a last resort, the tossing out of a desperate message in a bottle into the ether, one that washed into homes across the country and beyond in a last-ditch hope that a vital shared family moment or long-desired reunion might be facilitated before it was too late. In a sense the appeals were the shortest of short stories, telling you the bare minimum, leaving your imagination to fill in the rest. Indeed, by 1936 the messages had become such a part of the nation's cultural fabric that a collection of short stories based on them was published called *Missing from their Homes: Short Stories on the Familiar Broadcasts for Missing Persons*, a book whose contributors included Graham Greene, H.E. Bates, E.M. Delafield and Arthur Machen.

Yet from that cultural and societal pinnacle the SOS message slipped gradually away into radio history. No longer do we eavesdrop inadvertently on another family's anxiety: imminent death is easier to disseminate privately when we all have a phone in our pocket making us permanently contactable and an online digital footprint that makes us easily, even instantly traceable.

'I don't think I've read one since we moved to TV Centre in 1997, thinking about it,' Corrie Corfield had told me. 'The schedule changed that year too so the 17.54 shipping forecast went out on long wave and *PM* carried on on FM, which might have affected their broadcast. I certainly don't remember ever receiving a memo saying they'd been withdrawn officially, though. I think technology has just changed the way we interact with each other in that kind of situation.'

Occasionally newspapers would publish follow-up stories, many of them containing extraordinary tales of community-mindedness and of people willing to go well out of their way for a person they'd never met, motivated only by compassion. But as far as most listeners were concerned the appeals just vanished into the ether. So did anyone get hold of Mrs Carr of Flitwick and did she make it to the Middlesex Hospital to see her son?

Frank Carr was a 32-year-old widower when his worsening diabetes forced him to give up his job as head railway shunter in a London marshalling yard. On collapsing in April 1923 he was rushed to the Middlesex Hospital unconscious where he fell into a coma and was given a bleak prognosis. Frank's mother Rebecca was not on the telephone and lived in a remote part of Bedfordshire, leaving hospital staff flummoxed as to how they might contact her urgently. It was a hospital porter named Harold Carter who came up with the idea of calling the BBC and asking if they'd consider broadcasting an urgent appeal. A doctor named Bates, living eight miles from Flitwick and at home listening to his wireless, heard the appeal, jumped straight into his car, found his way to Thinnings, collected a bewildered Mrs Carr and drove her to Luton railway station in time to make the 10.15 p.m. train to King's Cross. There a taxi was waited to rush her to Fitzrovia and a little over two hours after the broadcast Mrs Carr arrived at her son's hospital bedside, where she remained until he died the next morning.

'If it had not been for the broadcast message I should probably not have reached my boy until it was too late, as it was through the broadcasting of the appeal for someone to fetch me speedily to the hospital that I made it in time,' Mrs Carr would recall later. 'I'm told that between 18 and 20 people called at my house with news of the wireless message.' With only around 30,000 radio-receiving licence holders in Britain at the time the announcement was a bit of a shot in the dark, but the number of people who responded proved that it was a shot surprisingly likely to hit its target.

Mr Carter, the hospital porter, retreated immediately back into the mists of history, but to an extent we have him to thank for the countless messages of mercy that would be disseminated via the airwaves in the following decades. Was it possible that Carter had heard the very first radio SOS, broadcast a month earlier?

It's not known who had the idea to put out a message on 15 March 1923 appealing for sightings of a boy scout who'd gone missing from his home in Birmingham, but a weekly Thursday night programme aimed specifically at boy scouts must have seemed exactly the right place to do it. Either way, Bristol police heard the message, made a note of the boy's description and, sure enough, recognised the young man who had somehow dib-dib-dibbed his way as far as Bristol only to be promptly dob-dob-dobbed straight back home to Birmingham, apparently none the worse for his adventure. 'Wireless has now a triumph to its credit in the detective line,' said the *Daily Herald*.

A month later Mrs Carr made her dash to Fitzrovia and the age of the SOS had begun.

So popular did they become that in 1933 the BBC had to introduce guidelines governing their nature, limiting the appeals to relatives of the dying and urgent police messages.

A doctor or appropriate medical attendant had to confirm the patient was indeed dangerously ill, and even then all other means of communication had to have been attempted without success. 'No SOS can be put out regarding animals or property,' the Corporation declared: owners of missing cats or forgotten umbrellas would have to make their own arrangements.

The accounts below all arise from genuine SOS messages broadcast on BBC radio that I've found by trawling old newspapers that followed up the stories. Many of them showcase the empathy inherent in the human condition. Some have genuinely happy endings, others tragic. Others remain open-ended, or are born of circumstances that leave more questions than answers. All-in-all these stories represent the peak of radio as a public service. Each of them is a little slice of social and domestic history, tiny dramas prompted by the desperate last call of a voice in the ether.

'Will Mrs Christina Shepherd, who may be known as Mrs Christina Dormer, last heard of 12 months ago at 2 Portland Terrace, Greenford, Middlesex, go to the Lake Hospital, Ashton-under-Lyne, where her daughter Elsie Shepherd is dangerously ill.' (6 February 1939)

The Dormers hadn't used their wireless for a while and when they switched it on during the early evening of 6 February it was only to check whether their mantelpiece clock was displaying the correct time. Instead, once the valves had warmed up, Christina Dormer heard her name announced followed by that of her 19-year-old daughter Elsie, from whom she hadn't heard in a year. Immediately the Dormers raced to the station in time for a night train north where a remarkable story was playing out.

At four o'clock that morning the door of a remote transport café at Woodhead, in the teeth of a blizzard high in the Pennines on the main road between Sheffield and Manchester, had banged open and in a flurry of snow a lorry driver called urgently for help. Draped around him was a heavily pregnant and barely conscious young woman in the early stages of labour. It was a filthy night whose extreme conditions had prompted café owner George Tharme to keep his premises open with the fire blazing in the grate as a potential haven for any lorry driver who might become stranded on the road in sub-zero temperatures.

While the drivers tried to make the stricken Elsie, who wore only a thin coat, as comfortable as possible in front of the fire, Tharme ran to his van and made the perilous drive a mile and a half down the valley to a pub with a telephone. Eventually he managed to reach a Dr Golding, who battled through the snow to arrive at the café just in time to deliver Elsie of a son. He then drove the newborn baby and the woman, who was very weak and calling for her mother, to the hospital at Ashton-under-Lyne, where the call was put into the BBC hoping to trace the girl's parents.

Elsie Shepherd had disappeared from the home she shared with her mother and stepfather a year earlier. When they were reunited in the hospital all Elsie would say when she recovered her strength was that she'd been travelling the country and had worked in a Manchester café for a while. Whatever the cause of her departure, her mother's arrival proved to be vital to Elsie's recovery. 'There is no doubt,' said a doctor, 'that her mother's presence saved the girl's strength at the crucial moment.'

'Would Miss Joyce Wilsher, currently on holiday in Carlisle, please go to 2 Wicklow Street, Old Basford, Nottingham, where her fiancé Raymond Holmes is dangerously ill.' (15 November 1952)

'If I could see her I could make everything right, I'm sure.' Raymond Holmes was nothing if not optimistic. He was also far from dangerously ill, unless you counted the broken heart that drove him to take desperate and ill-advised action designed to provoke the sympathy of his erstwhile sweetheart. Raymond and Joyce had been engaged to be married for two years until Joyce called the whole thing off a week before the SOS was broadcast, an action that devastated Raymond. In order to give herself a bit of peace from his constant pleading to reconsider, Joyce told Raymond she was going to Cumbria for a while to clear her head. In actual fact she remained at home with her parents, a few streets away from the house where Raymond lodged and where she heard the message broadcast on the BBC radio news as she helped her mother prepare lunch. Fearing that a bereft Raymond, deciding he couldn't live without her, had tried to kill himself she rushed to his door along with her brother Harry only to find her apparently gravely ill ex-fiancé had just popped out.

That morning, determined to track down Joyce, Raymond had telephoned the Carlisle police. Whatever cock-and-bull story he gave them it prompted them to suggest he try a radio SOS via the BBC. In line with the Corporation guidelines his GP confirmed in good faith that Raymond was in a very fragile state of health and the message was read out on the air.

When Raymond returned home to learn that Joyce had called at the house within minutes of the broadcast to find him not at home he was aghast. He ran immediately to the Wilsher home where Joyce's mother refused to let him through the front door. It was there that he tried to convince Mrs Wilsher that if only he could see Joyce he would make everything all right. Mrs Wilsher was having absolutely none of it.

'Raymond is a nervous type of youth and had been receiving treatment for his nervous trouble,' said his landlady, Mrs Oakland. In the days following the incident the BBC was forced to issue a statement. 'We obtained confirmation from the doctor handling the case and since we have obtained a second confirmation,' it said. 'So far as we are concerned that is the end of the matter.'

The following year Joyce Wilsher married a man named Wilfred. Raymond never married.

'Would Mr George Arthur Vowles, on a cycling tour in the east of England, return home to Buxton where his wife is dangerously ill.' (3 August 1936)

When 49-year-old clerk George Vowles set off from the home in Buxton he'd shared for 21 years with his wife Margaret, he was looking forward very much to a few days cycling the flatlands of east Yorkshire and Lincolnshire. Margaret was 15 years older than George, and at 64 her cycling days were behind her, but she told him to send regular postcards to keep her abreast of his movements. The day after George set out in fawn plus-fours, brown golf jacket and navy beret Margaret had gone to visit a friend in nearby Fairfield Road for tea. After complaining of feeling a little off-colour Margaret collapsed unconscious and a doctor was called. The prognosis was grave and he asked after Margaret's husband. When told he was on a cycling holiday the doctor said it could be a matter of hours and George should be there.

When the SOS message was read on Monday 3 August an oblivious George had just set out from Stockton-on-Tees. The previous evening, as his wife lay gravely ill in Buxton, he'd sent her a postcard telling her he planned to travel home via Whitby, Scarborough, Boston and Chesterfield in time to be back at work the following Monday.

Margaret held on for a couple of days but died on Wednesday 5 August. A family friend Francis McDearmid set out in his car to try to find George, without success. The postcards kept coming. On the Tuesday he'd been at Grantham and was setting off for Cleethorpes. On the Thursday it had 'turned out wet this afternoon and I had to go to the pictures. I don't know what to do'. The last postcard arrived on the Saturday, postmarked Swaffham, Norfolk. 'Many happy returns for Sunday!' it read. 'Had a puncture, landed here at 1.40 a.m. Going on to Peterborough and then for home.'

'Missing from his home at Stoneferry, Hull, since 13 June, Clarence Farnes, also known as Render, aged 24 years, 5 foot 6 inches in height, slim build, fair hair, grey eyes, fresh complexion, has the end of the middle third and fourth fingers of his left hand missing, dressed in a brown suit, white striped shirt, collar and brown tie, blue socks, back boots and bowler hat. This man suffers from neurasthenia and may be wandering due to loss of memory. Any information should be given to Hull city police, telephone Hull 2690.' (7 August 1933)

Clarence Farnes seemed a little preoccupied when he came home for his dinner on the warm, sunny evening of 13 June after a day selling fruit from his barrow around the streets of Hull. It was a decent job, one the 24-year-old was more suited to than his previous employment as a sawyer, where frequent accidents had cost him two of his fingertips, and he didn't seem to have too much to worry about. He was a little quiet over his meal, but his mother Edith recalled no reason for concern. He went out after dinner without saying a word and didn't come back. June became July, and when July became August Edith turned to the BBC.

On 31 August, three weeks after the broadcast which had produced no sightings, friends of Clarence in Cottingham, some 40 miles away, answered the door to a filthy man with wild hair, a full beard and a badly sunburned face. 'Tell Mother and Father I'm back, they'll be worried,' he said, and promptly collapsed.

Clarence Farnes couldn't tell anyone much about his ten-week absence. When he arrived in Cottingham the boots he was wearing were not his, and nor were the trousers, which had been torn off at the knees. When his parents had collected him and taken him home it took three baths before he was clean enough to be put to bed.

'I should hardly have recognised my own boy,' said Edith.

'I fell and hit my head,' Clarence said of his return. 'The blow seemed to shake me up and my memory, which had been blank, came back and I realised I was miles from home.'

'Missing from her home at Magdalen College, Oxford, since Friday 21 April, Miss Edith Margaret Kennard Davies, age 21, 5 foot 7 inches in height, brown bobbed hair, hazel eyes, fresh complexion, probably dressed in brown skirt, brown suede jacket and stockings, no hat. This woman, who is in possession of a green enamelled bicycle, has recently recovered from illness and may be suffering from a loss of memory.' (28 April 1933)

Edith, the daughter of the headmaster at Magdalen College School, had set out on her bicycle to visit Dr Good, the medical superintendent at the Littlemore asylum. The doctor had treated Edith for insomnia and a loss of appetite when she was anxious about forthcoming examinations, but she'd recently returned from a hiking holiday with her sister, apparently in good spirits. Her visit to Dr Good that day was to ask his advice about a possible nursing career, but she never arrived. Her mother asked the BBC for help but was told that a person had to have been missing for seven days before they could broadcast an appeal.

'That is a long time to anyone who is anxious,' said Mrs Davies. 'No doubt my daughter is behaving perfectly rationally but she may be unaware of her identity.'

On 2 May a 13-year-old boy noticed a green bicycle lying by the side of the road close to Mastells Wood in Shotover, outside Oxford. When it was still there the next day he told his father, who alerted the police. A search party gathered, including Edith's brother Arthur, a merchant seaman, who found Edith's body lying in the undergrowth well away from the footpath. Arthur collapsed and had to be carried from the wood to a waiting car. Edith lay in a natural position with her clothes undisturbed, a handbasket and pair of gloves placed nearby.

The inquest revealed that Edith had consumed six sticks of metaldehyde, a fuel for camping stoves that if consumed in sufficient quantities induced convulsions, coma and eventually death. The inquest concluded that Edith had died as a result of consuming the fuel sticks, but there was 'no evidence to show whether she knew the substance was poisonous'.

'Missing from her home at Albion Cottage, Sandhurst, Berkshire, since 13 August 1929, Winifred Parrant, a domestic servant aged

20, 5 foot 4 inches in height, slim build, sallow complexion, brown eyes and dark brown hair with fringe, dressed in a navy blue coat and shirt, blue and white striped jumper or mauve jumper, flesh-coloured stockings, black one-bar shoes, inclined to limp on left leg. Last seen near Virginia Water, Surrey. Will anyone able to give any information as to the whereabouts of the young woman please communicate with the chief constable, Surrey County Constabulary, telephone Guildford 83, or any police station.'
(11 January 1931)

When Private T.H. Jennings of the 2nd North Staffordshire regiment confessed to the murder of Winifred Parrant it seemed the mystery of her disappearance had been solved at last. In early January 1931 Jennings, who had been posted as a deserter, told police he had been in a relationship with Winifred and killed her in a jealous rage close to a golf course in Virginia Water, Surrey. He added that her body was hidden in gorse bushes in Windsor Great Park, while a bloodstained army uniform found in Woking that contained letters signed 'W.P.' seemed to corroborate the account. The police began to search the park.

Jennings said that he had met Winifred at the pictures in Camberley and they had fallen in love, only for another soldier to arrive on the scene and turn her head. Jennings discovered this on 11 August 1929, two days before Winifred was reported missing, when they were on their way to visit Winifred's sister in Brighton. He confessed to police that he had killed her by kneeling on her throat.

When the search in Windsor Great Park turned up a woman's hat and an attaché case containing underwear it seemed the police were edging closer to finding Winifred's body. Her family waited anxiously for news: in the weeks following Winifred's disappearance her sister had received a letter, apparently from Winifred, claiming she was on her way to Canada, while a letter to her grandfather said, 'I am quite all right and in a good job. I could not stop at home. I am very sorry for all the trouble I have given you. Don't try and find me.'

The newspapers were full of the soldier's sensational confession and the Windsor Great Park dig, which is when Winifred Parrant, now

Mrs Richard Jenkins and the mother of a five-month-old daughter living in Greenwich, south-east London, happened to read the story of her own murder over the shoulder of a fellow passenger on a tram. Winifred's story was a little different from the alleged killer's. She had never met Private Jennings, she said. The hat and attaché case found in the park were not hers. 'There in big bold type I saw the name Winifred Parrant and at first thought it was a coincidence. When I got off the tram I bought a paper and read the whole story,' Winifred said later. 'This was the first knowledge I had that the police and relatives thought me dead.'

Winifred said she had met Dick Jenkins, a plasterer by trade and a member of the Territorial Army, when he rode up to her on his motorcycle while training at Aldershot on 13 August 1929, the date she had been reported as missing. The couple fell immediately in love and married at Greenwich register office on 1 October that year.

'My impression of Winifred was that she was a thoroughly lovable, adventurous girl who had made up her mind to leave home and liked to be in London,' said Jenkins' sister. The couple had married in secret before going to Jenkins' parents' house where he introduced his new wife to the family. They had then lived quietly in Greenwich and in the summer of 1930 had a daughter. She was a poorly child who spent her early months in hospital, and Winifred blamed preoccupation with her baby for her complete ignorance of the news of her alleged murder that dominated the newspapers. 'When I read the paper I was too bewildered to know what to do,' she said. 'So my husband and I decided we should get away for a day or two.'

The Jenkinses went to stay with friends in Harlesden, north-west London, where after a few days Winifred went to the police station and nervously admitted that she was the woman whose strangled body they were looking for in Windsor Great Park.

'The Surrey police feel that, with the great assistance rendered by the press and the BBC, they have accomplished the task of finding Miss Parrant alive and no more credence is being placed on the confession made by Private Jennings,' read a statement issued from Guildford police station. When Jennings was told that Winifred had turned up alive and married, he reportedly collapsed in hysterical laughter.

'Will anyone in London or the south Midland area who on Saturday 19th December supplied a man with sandwiches containing sausage meat or some similar meat, which he took away wrapped in paper, communicate immediately with the chief constable of Oxford, telephone Oxford 3105. The man was probably riding a bicycle.' (20 December 1931)

One of the more unusual appeals for help in catching a criminal was this message regarding a sausage sandwich broadcast just before Christmas in 1931. The previous day 56-year-old Mabel Matthews was cycling home to Burford in Oxfordshire after helping a friend cope with the Christmas rush at her health food shop when she was violently attacked. A pair of cyclists found her lying bleeding by the roadside. Barely conscious she murmured the words 'brandy' and 'cyclist' before passing out and dying a few hours later. The coroner's inquest determined that Mrs Matthews had suffered 23 injuries, including 11 broken ribs and a fractured jaw. Her handbag was found half a mile away, while a man's bloodstained mackintosh was also found close to the scene, in the pocket of which was a sausage sandwich wrapped in paper.

Joseph Maddams heard the broadcast at his home in Northolt, Middlesex, and remembered the sausage sandwiches he'd prepared for his 22-year-old brother-in-law George Pople, a private in the South Wales Borderers, the previous morning before he continued the cycling tour he was making that would most likely take him through Oxfordshire on his way back to his regiment in Brecon. He called the Oxfordshire police and Pople was arrested the next day in Abergavenny. Although he denied murdering the farmer's wife a jury took barely half an hour to reach a guilty verdict, and George Pople was executed at Oxford Prison on 2 March.

17

Norman Conquest: Leonard Plugge and the Birth of Independent Radio

At the northern tip of the Place Charles de Gaulle, in the town of Fécamp on the Normandy coast is a branch of the Société Générale bank. It's a soulless-looking spot, the ground floor of the old building having been ripped out completely and replaced at some point – the seventies by the looks of it – with a pretty ghastly concrete and glass frontage. The sign over the door is just a red and black square with no name, as if the bank itself is ashamed to admit to its own premises.

As notable sites in the story of British radio go, this is about as unpromising as it gets. It's not even in the right country, for a start. Yet on a foggy August morning in 1931 this building played host to a chance conversation during a random visit that would change the face of British broadcasting. Independent radio launched officially in Britain as late as 1973 due to the combined intransigence of the government and the BBC to retain the monopoly of the national broadcaster, but its roots lie firmly embedded in a little town nestling among the cliffs of Normandy, and one remarkable man who one day happened to pass through it.

The previous spring Captain Leonard Plugge had founded the International Broadcasting Company, leasing airtime on European radio stations on behalf of commercial clients, with premises on Hallam Street in London, a hop and a skip from where the BBC's new base at Broadcasting House was under construction. Plugge was already a well-known radio enthusiast, having made a number of extensive car journeys through Europe in the 1920s on which he carried an elaborate set of radio apparatus and, in 1925, had brokered a deal by which Selfridges sponsored a 15-minute broadcast from the station at the top of the Eiffel Tower. The birth of the IBC was the boldest assertion yet of his determination to break or at least circumvent the BBC's broadcast monopoly in Britain, and that morning in Fécamp the penny dropped as to exactly how he might properly pull it off.

Before it was a bank the building on Place Charles de Gaulle was a well-known Fécamp institution called the Café des Colonnes. It took its name from the building's frontage, into which were built six classical-style columns salvaged from a local church destroyed during the French Revolution. Before it was the Place Charles de Gaulle the square was the Place Thiers, named after the statesman Adolphe Thiers, first President of the Third Republic.

Plugge had arrived by the early ferry to Dieppe, heading for Deauville. The sun was taking its time in burning off the morning mist making it a chilly morning, and by the time he reached the busy fishing port of Fécamp Plugge was ready for a cup of something hot. Spotting the Café des Colonnes on driving into the Place Thiers he pulled up outside and eased into a banquette seat at a small table by the window. Plugge had passed through Fécamp before but never stopped, and when the owner wandered over to him, plunging his hands into the pocket of his apron for pad and pencil, Plugge looked out at the busy square and asked him about the town.

Salted fish, mainly, came the reply. Fécamp was France's busiest port for cod, caught as far away as the tip of Greenland, Plugge learned, with the long-range fleet usually gone for six months of the year. There was also the Bénédictine distillery nearby, where a man named Alexandre Legrand had revived an old recipe for a liqueur that had once been produced by the monks of the local abbey and made himself a very rich man as a result. He took Plugge's order, turned towards the kitchen and then paused. 'Oh, and we have a radio transmitting station here,' he said. 'Radio Fécamp.'

Plugge turned his head away from the square and looked straight at the café owner. 'Oh yes?' he said.

'Yes indeed. It broadcasts in the evening, from the Legrand mansion. It's very popular here.' He indicated the square with his pencil. 'I have a friend over there who is a shoemaker and a couple of weeks ago Monsieur Legrand mentioned him in a broadcast and he was swamped with new orders afterwards.'

Within minutes Plugge was knocking at the door of the Legrand residence, an elaborate three-storey villa opposite the distillery, and introducing himself to Fernand Legrand, the man behind Radio Fécamp. Legrand, a wireless enthusiast of long standing, was well aware of Plugge and his work and ushered him inside with considerable excitement.

Legrand had had a wireless licence since 1926 and had been president of the local radio club ever since. What excited Plugge the most was that Legrand had set up two radio masts on the cliffs a couple of hundred yards away, which gave him a broadcast reach capable of servicing the major towns of Le Havre, Rouen and Dieppe – and, Plugge realised, probably Britain. Before Legrand had even shown Plugge his transmitting equipment – behind the grand piano on the other side of the room – Plugge had struck a deal by which he would buy airtime on Legrand's station at a rate of 200 francs per hour, starting the following Sunday when the BBC would be putting out its regular dreary Sabbath fare of classical music interspersed with po-faced monologues on matters of moral import.

Returning to his car Plugge waved at the owner of the Café des Colonnes as he hopped into the driver's seat, pointed the vehicle towards Le Havre and went in search of the town's English bank, for he had gramophone records to buy and needed to withdraw some cash to do it. He was served by a 32-year-old Englishman named William Evelyn Kingwell, a man whose bank inspector father had encouraged a career in banking for his eldest son from an early age and who was working his way steadily up the career ladder at Lloyd's. The one time he'd been away from banking, fighting in the First World War, he'd been invalided home after being gassed and returned to the safety and security of the banking industry once he'd recovered. Kingwell was in Le Havre on secondment from London and always enjoyed talking to the Brits who came through the door. Plugge was understandably animated about his new project and told Kingwell all about it, asking if he knew of any Englishmen in the area who might be willing to go to the Legrand residence in Fécamp every Sunday night and play gramophone records over the ether.

Kingwell thought about people he knew who might fit the bill. Then he thought about his constant round of credits and debits, overdrafts and mortgages, the walk to work from his lodgings carrying his tightly furled umbrella and briefcase, placing his bowler hat on the same hook of the hat stand every morning and taking it off again every evening, just as he'd done every day since he left school and joined the Stroud Green branch of Lloyd's exactly half his lifetime ago. While he'd been pleased about his secondment to France the Le Havre branch was in effect no different to any branch in England. The same bowler hats, the same umbrellas, the same counters, the same filing systems. He looked back at Plugge, saw the animation and excitement in his expression, thought for a moment, took a breath and said he'd rather like to give it a go himself, if Captain Plugge didn't mind.

Captain Plugge didn't mind in the slightest. Having secured his presenter, he trousered his cash, spent an hour buying just about every gramophone record in Le Havre, delivered the box of shellac back to Kingwell at the bank, scribbled down a few notes as to what he had in mind, drew a map of how to find Legrand's house in Fécamp, clapped Kingwell on the shoulder, wished him luck, raised his hat and swept out through the door with a swish of his raincoat in a breeze of new purpose.

On the evening of 6 September 1931, as the wireless listeners of BBC Britain sat down to a church service live from St Leonards-on-Sea, a report from the London Clinic for the Treatment and Study of Nervous and Delicate Children and a concert of Wagner and Mozart from the BBC Orchestra, William Evelyn Kingwell, bank teller, motorcycled the 25 miles from Le Havre to Fécamp very carefully with a box of gramophone records strapped to the back of his bike, introduced himself to Fernand Legrand, was given a quick lesson in using the equipment and, once the BBC had ended its evening's broadcast with *The Epilogue*, a Bible-reading that closed its Sunday programming, sat at a table by the grand piano in Legrand's living room, pulled the gramophone player a little closer to him, cleared his throat, opened up the transmitters and became the first British disc-jockey to broadcast to Britain from a commercial radio station. For the next two and a half hours he passed the microphone between his lips and the horn of the gramophone, probably not realising the significance of what he saw merely as quite a nice break from routine.

Meanwhile, back in London, Plugge had been trying to secure publicity for his new venture. He also needed to secure the services of a more experienced radio voice and, importantly, find commercial sponsors for his Sunday night slot. The potential for the Fécamp station was enormous, he felt: not everyone shared John Reith's Presbyterian determination to keep the Sabbath a fun-free zone. Nobody necessarily wanted the airwaves filled with bacchanalian excess either, but something in between, that was actually entertaining to listen to, wouldn't go amiss. Just having a choice would be nice. Being hectored by droning wispy-haired clergymen when actually in church was one thing, enduring the same damning judgements of one's personal morality in one's own living room was another altogether.

Despite the clear potential of the project, however, Plugge's first press release had fallen largely on stony ground. The only flicker of interest came from the *Sunday Referee*, where Plugge's missive had landed on the desk of the paper's reporter with a radio brief, Stephen Williams, who was possibly the only man in Britain with any kind of claim to a commercial broadcasting background.

In 1928 the *Daily Mail* had launched a crass stunt in an effort to repeat the success of its sponsorship of the Nellie Melba recital eight years earlier. It had chartered a cruiser called the *Ceto* at Dundee, loaded it with amplifiers and enormous speaker horns and packed it off around the coast of Britain for the summer, anchoring off seaside resorts to bombard the beaches with gramophone records, relentless promotion of the *Mail*, its sister papers and an insurance policy the paper was flogging at the time. There were speeches from local dignitaries and even local 'turns' rowing out to the *Ceto* to do their thing at ear-splitting volume for the aural pleasure of the bathing public. At night the ship sparked up two searchlights to illuminate whichever 'prom' it was moored off and the words '*Daily Mail*' were picked out in 1500 red and white lightbulbs on the hull. Imagine having that inflicted on you as you rolled up your trouser legs for a paddle. We're not talking top-quality Dolby sound, either, rather enormous amplifier horns sending out a tinny, distorted racket across the breakers accompanied by the screams of gulls and the yelling of terrified children wondering why their ears were hurting. If it turned out to be even half as horrendous as it sounds you'd think it would have cleared every beach in a heartbeat and been chased off by a flotilla of harpoon-wielding locals, but amazingly the experiment was deemed a success, albeit one that was never repeated.

The man in charge of the *Mail*'s microphone that summer was Stephen Williams, who'd written to the paper as a student looking for a job during his summer holidays from Cambridge University, and ended up skirting the nation's shores shouting at holidaymakers about insurance through a PA that sounded like a comb and paper would if put through Pink Floyd's touring rig.

Williams loved Plugge's plan and agreed to print the station's frequency and timings as 'Special Broadcasts for British Listeners' for an initial three weeks. So delighted were wireless listeners to discover an alternative to the stern Reithian chill of a wireless Sunday they not only tuned in in droves, they sent the *Sunday Referee*'s circulation spiralling upwards into the bargain. It was rough, it was seat-of-the-pants stuff, it was basically a career bank clerk working his way through a pile of hastily compiled 78s in a Frenchman's living room and reading out the labels, but this is where Capital Radio, Absolute, Heart, Classic FM, Talksport, Virgin, Radio X, the pirate stations, all of them were born. And it was all down to Leonard Plugge.

The son of a commercial traveller from the Netherlands, Leonard Plugge would lead a life of adventure, innovation, wealth, poverty, joy and tragedy. Although he was born and spent his early years in south London Plugge grew up in Belgium, showing an early aptitude for ice-skating that saw him crowned male ice-dance champion of Brussels and a pairing with future three-times Olympic gold winner and Hollywood superstar Sonja Henie. The outbreak of the First World War saw the 25-year-old Plugge move back to England in 1914 where he completed an engineering degree as a refugee student at University College London, joined the Royal Flying Corps and, despite being frankly terrified of flying, was made captain shortly before the end of the war. He worked in Berlin for the Allied Aeronautical Commission after the Armistice, becoming the youngest Fellow of the Royal Aeronautical Society in the organisation's history. In 1922 he swapped the open skies for the tunnels beneath London, transferring his engineering skills to the statistical department of the London Underground where he met a wireless enthusiast named Bill Wood. Plugge fell immediately in love with the medium of radio when Wood invited him round to listen to the new BBC Birmingham station one evening. So obviously smitten was he that Wood let him have the set he'd built, making a new one for himself. Before long Plugge found himself on the committee of the Radio Society of Great Britain.

Given his pan-European background it's no surprise that Plugge's interest in the wireless stretched beyond the shores of the UK. He'd sit up late into the evening, awestruck as he picked up stations from as far away as Romania or the Azores until, in 1924, he decided to pack up his radio, cross the Channel and visit some of these exotic places himself. Licensing regulations meant he needed a permit to listen in each country visited and he had his set confiscated briefly on his first train journey between Paris and Zurich, but he spent a successful couple of weeks trundling around Mitteleuropa with his radio, setting it up in hotel dining rooms, unravelling its 100 feet of aerial wire and picking up the Savoy Orpheans dance band from the Strand in places as far afield as Italy, France and Austria.

Plugge's late-night listening to the European stations, and the people he'd met when visiting them on his radio InterRail, convinced him that the British public – with their domestic choices limited to a BBC whose promise to inform, educate and entertain was scoring only two out of three – were ready for a new radio experience. He began to think about how he might open up Europe and its wavelengths to British listeners. Although still ostensibly an employee of the London Underground Plugge threw himself further into his radio activities later in 1925 when he set out on another European tour, this time by car. He adapted his Paige Tourer to accommodate a radio set with a frame aerial made from wire and wood affixed to the outside of the bodywork, a tuning device next to the steering column and five sets of headphones, one for Plugge and four for the paying passengers required to help fund the trip: four women schoolteachers who answered an advert Plugge had placed with a travel agent. Named 'Aether 1' the Tourer was hoisted onto the cross-Channel ferry at Newhaven, hoisted off again at Dieppe, and the adventure began. Plugge reported that he could still receive the BBC's 2LO London station as far south as the Basque country, where he could pick up opera from Covent Garden loud and clear thanks to the company's powerful new transmitter at Daventry.

Reaching Paris Plugge went to see Maurice Privat, who ran a new station broadcasting from the Eiffel Tower using equipment installed by the military during the First World War. Plugge had met Privat on his trip the previous year and the two men had discussed the possibility of a sponsored broadcast in English. Privat had been agreeable and Plugge had gone back to England to convince Selfridges. Hence on 29 August Plugge sat down at the microphone himself and held forth for a quarter of an hour on the magnificent range and quality of Selfridges' women's fashions, ending the broadcast asking anyone in Britain who'd heard the brief programme to write and tell him so. Having had no opportunity to publicise the event Plugge later said that he heard from only three people who had happened across the broadcast by chance.

While the first ever commercial broadcast in the English language may not have been a commercial success it only further fired Plugge's enthusiasm for the possibilities of radio competition for the BBC. In the autumn of 1926 he planned an even bigger European jaunt, in an even bigger car with an even bigger radio set installed. Motoring in a six-seater Packard fitted with a specially adapted Western Electric receiver he set out for Constantinople accompanied by five women whom posterity records only as Dorothy, Ethel, Florence, Mildred and Myfanwy (Plugge, it seemed, enjoyed the company of women on his tours). It was by far his most ambitious trip yet, and with Maurice Privat promising to deliver nightly progress reports on his Eiffel Tower station, the publicity potential was ratcheted up like never before. Indeed, with advance notice of their route and Privat's bulletins the motley collection of Brits in their giant car found themselves greeted by crowds that increased in size the further south-east they ventured. The demands of the crowd and some rudimentary road surfaces meant that instead of reaching Turkey Plugge was forced to turn around in Bulgaria and start the journey home, but not before tuning in to the midnight chimes of Big Ben and having them ringing out in a Sofia street from 1400 miles away. The return journey saw them chased through a Balkan village after accidentally running over a popular goose, and frantically pushing the car out of the way of an oncoming train after stalling on a railway crossing near Zagreb, but they arrived safely in London after a trip of 4,500 miles that had taken five weeks.

Plugge made another trip the following year, to Lisbon, with another three women, one of whom was summoned home by a BBC SOS picked up in Portugal after her father was taken ill. Plugge's adventures finally persuaded him to jack in his job at the Underground and pitch in full-time with his new IBC venture. (It's possible that the Underground's reaction was, 'You're leaving, Plugge? We didn't even know you still worked here.')

While Kingwell was still faithfully turning out the weekly show from Fécamp, back in London Plugge had identified his permanent replacement: Major Max Stanniforth, an ex-military man who'd worked on the South American railway system and happened to be looking for a new challenge when he spotted Plugge's advertisement in the *Daily Telegraph*. Plugge was a man impressed by military titles, and took on Stanniforth at a salary of £4 a week. It was a timely appointment: the novelty of the late Sunday nights before an early Monday start at the bank and the bumpy two-hour round trip by motorcycle was starting to wear thin for Kingwell, not to mention lingering health problems from his wartime gassing sometimes making it difficult for him to speak for the entire two-and-half-hour programme. At the end of December Stanniforth arrived and a relieved William Evelyn Kingwell puttered back into the mists of radio history on his faithful motorcycle.

At the behest of the *Sunday Referee* the station's transmitter power was dramatically increased and the name of the station changed to the less parochial Radio Normandy. Great efforts were undertaken to make the operation sound much larger than it was. Relocating to a barn above the distillery stables hung with horse blankets to deaden the sound, and stewing in the heat of a fearsome wood-burning stove, Stanniforth and Stuart Williams, who had also joined the station as a presenter, would pointedly refer to Normandy's other studios and even play recordings of local dance bands claiming they were live recitals by 'the IBC Band'. The station was still largely experimental at this stage but growing fast. When a listeners' club was started in 1932 it received 50,000 membership forms on its first day, a number that would soon grow to half a million. Before long the late-night Sunday broadcasts had expanded to encompass the entire week. The key measure of success, however, was whether there was money to be made. Fécamp was no Writtle, bankrolled by a huge company as a research exercise. There needed to be a bottom line, or the whole venture was pointless. One of Plugge's executives at the IBC in London, an old university friend named George Shanks, came up with the idea of devising a women's facial cream from a recipe he'd seen in a book belonging to his mother. He made up a few sample pots in his mother's kitchen and sent them over to Fécamp under the name Remis, produced by a fictitious company named Classic

Beauty Productions. Between them Stanniforth and Williams came up with a long advertising spiel for Remis that evoked the great beauties of classical antiquity and gave Sharp's mother's London home address at the end of the pitch. Within days Mrs Sharp was being swamped by orders pouring in from all over the country. If a potion hastily concocted in a saucepan and punted by a couple of daft radio executives losing the run of themselves could sell like that, the money to be made by advertising real products was clearly something quite substantial.

The BBC was far from happy at the rise of Radio Normandy, which was comfortably outdoing them in listener numbers on a Sunday despite transmissions only reaching as far north as Yorkshire and north Wales. But there wasn't a great deal it could do beyond impotent appeals to the government and a terse reminder on the licences it issued to listeners that the licence was only valid for 'duly authorised broadcasting stations'. In 1933 questions were asked in the House of Commons regarding British goods being advertised in English from foreign stations, and an official protest was sent to the French government from Westminster, but unsurprisingly nothing came of it. Reith wouldn't concede much on Sunday broadcasting, turning the day of rest into the battlefield on which Plugge started the fight for independent commercial radio against the BBC monopoly. It took until the Sound Broadcasting Act of 1972 for legal independent radio to arrive, the first legal, UK-based commercial stations launching the following year thanks largely to the success of the offshore pirate stations of the 1960s, but Plugge did at least live to see independent commercial radio legalised, even if man had landed on the moon by the time it happened.

While Reith's religious intransigence can shoulder much of the blame for the growth of a kind of radio he abhorred, most of the credit is due to Leonard Plugge. The IBC went from strength to strength during the 1930s, launching the careers of broadcasting legends like Roy Plomley and Bob Danvers-Walker, among others, and buying into a range of continental stations such as Radio Paris, Toulouse, Athlone, Ljubljana, Valencia and Rome, allowing Plugge the wealth and time to expand his activities, ambitions and eccentricities. One IBC employee recalled that at one point Plugge had 43 telephones in his London flat, including five by the bath, all in different colours. It paid off for the companies advertising with Plugge too. One such was the technical retailer Philco, which became a household name; another, Hanley's car sales company, opened a nationwide chain of showrooms on the back of their radio advertising success.

In 1934 in New York, after a whirlwind romance, he married Ann Muckleston, an American model and actress 20 years his junior, before the following year being elected to Parliament as the Conservative candidate for Chatham under the slogan 'Plugge in for Chatham'. Until that point Plugge had spent a lifetime patiently pointing out that his Flemish surname was pronounced '*Plooje*' but his run for Parliament meant that after a lifetime of 'actually it's...' he finally had to concede pronunciational defeat. Once elected to the Commons he managed to flaunt both his eccentricity and his wealth by purchasing a 110-foot motor yacht that he christened the *Lennyann*, in which he would sail up and down the Thames between his constituency and the Houses of Parliament.

The Second World War signalled a drastic change in Plugge's fortunes and he would never recapture his astonishing pre-war success. The European stations that provided him with his whopping personal income were either commandeered or closed down altogether at the start of the conflict. Plugge lobbied unsuccessfully to be put in charge of a Forces broadcasting network, and when he sent his wife and young son to America for the duration of the hostilities, Ann began a relationship with another man and asked for a divorce. Rushing to America to win her back Plugge suffered a mild heart attack but still succeeded in bringing his wife and son home, with Ann giving birth to twins the following year.

Radio Normandy never re-opened and Plugge lost his parliamentary seat in the Labour landslide at the 1945 General Election. He kept himself busy with the IBC and interests in Radio Paris, Radio Toulouse, Radio Ljubljana, Radio Athlone and a number of Spanish stations, and on a visit to the USA bought an enormous Buick that he had shipped over to London and would take great delight in parking outside Broadcasting House when he visited the IBC offices in Portland Place – but what victories he scored were becoming pyrrhic at best.

He sold the IBC in 1964 and decamped to Rome in order to study painting and sculpture at the Accademia di Belle Arti, spent a year in Florence and then returned to Rome where he became a familiar figure in the streets dressed in an artist's smock and Rembrandt-style cap. In 1966 he returned to London and his flat in Dolphin Square, Pimlico, which he'd filled with classical sculptures and decoration, and steadily began to run out of money. As the 1970s dawned he was so broke that the electricity was cut off at Dolphin Square, leading to a bad fire when a candle set light to the curtains. Soon afterwards he relocated to California.

Plugge's final years were blighted by tragedy. In the early seventies his daughter Gale Ann became involved with a Black Power activist named Hakim Jamal, an associate of the controversial British-based Michael X, and she moved with Jamal to the commune set up by X in Trinidad after he'd skipped bail in the UK on charges of robbery and extortion. In January 1972 Gale Ann was set upon and buried alive by some of Michael X's followers, her body being discovered a few weeks later. According to some sources Michael X had ordered the murder because he felt Plugge's daughter was too much of a distraction to Jamal. A year later Gale Ann's twin brother Greville was killed in a car crash in Morocco, leaving Plugge bereaved as well as nearly broke. He spent his final years hawking an invention by which a car would be fitted with two batteries and a switch to transfer power between them but it never caught on. There was one last flirt with the spotlight when, at the age of 90, he gave a gloriously eccentric television interview to Alan Whicker in 1980, the last hurrah of an extraordinary man before he died after a stroke early the following year.

Leonard Plugge is no longer a famous name, but he was a truly innovative radio pioneer. That chance encounter in a Fécamp café on a chilly August morning in 1931 started a process that changed the entire face of British radio. It would take more than four decades and require the growth of pop music and the shipboard pirate stations that played them, but even they owed a debt of gratitude to Leonard Plugge. He lived a remarkable life largely on his own terms, even though his luck seemed to desert him from the moment war broke out and French forces sabotaged the Fécamp transmitter ahead of the German invasion. From his early European vehicular odysseys – in which he practically invented the car radio – to having the 1970 Mick Jagger film *Performance* filmed in his London home, to losing two of his children in heartbreaking circumstances, Leonard Plugge's was a remarkable radio life.

There is one lasting legacy of the man who effectively invented British commercial broadcasting. That 1925 Selfridges broadcast from high up in the Eiffel Tower, where he'd spent a quarter of an hour talking up the exclusive London store's range of women's fashions, was the first of its kind. It happens everywhere now, and every time it does there's a small tribute to the man who did it first. Not for nothing is it known as a 'plug'.

18

Not Fit for Human Habitation: Two Lochs Radio, Britain's Smallest Commercial Station

Gairloch is a very long way away indeed. It doesn't matter where you might be reading this, Gairloch is, I guarantee, a very long way from wherever that might be. If you're driving from the south – and I'd be prepared to wager my favourite trousers that you would be, because there are only a few who would be approaching from the north – you have to go to Inverness and turn left. This is the last leg of the journey, the home straight, the most direct route, and it literally involves covering the entire width of Scotland. In travelling between Inverness and Gairloch you couldn't go any further in either direction without getting your feet wet.

It's worth it, though. For one thing, that last stretch of the journey is a beautiful drive, especially on a crisp winter's day when the sun rises just far enough into a cloudless blue sky to turn the air itself a rich gold and paint the snow-sugared whiteness of mountains with a thin golden veneer. For another, it's a place designed for remote contemplation. A long, thin town, Gairloch is a sprinkle of buildings along the wide shore of Strath Bay that look out across Loch Gairloch to layers of peaks beyond. There are few places I've ever been to in Britain and Ireland where the rest of the world feels as agreeably far away as it does in Gairloch.

Take John Macleod, for example. In 1841 he sailed alone into the harbour at Gairloch, tied up his yacht and never left. He fished at first, kept himself to himself, but engaged enough with the locals to attract the attention of the laird who, when Macleod's boat became damaged beyond repair, gave him a job on his estate. Macleod worked quietly and diligently and after a few years was appointed the estate manager, combining that role with being the agent for a steamer company. He was an intelligent man with a good head for business, was kind to the poor, could discuss English and Scottish literature knowledgably and followed horse racing keenly. The only faintly odd thing about him was a reluctance to talk about his past.

In 1866, 25 years after his arrival in Gairloch he died, 'much regretted by all who knew him' according to a contemporary account. It was only after his death that the story of 'John Macleod' emerged, told to a friend by the man himself as he lay on his deathbed. On the other side of Scotland a young man had farmed a smallholding and worked in a local bank, where, reliable and trustworthy, he had risen to become principal teller. Then one day he didn't arrive for work and was never seen again. A couple of weeks after his disappearance the bank uncovered a fraud on a massive scale, one that must have taken intense planning and been exercised over a long period of time and one that netted the perpetrator a fair pile of cash. The police were called in and despite his fortnight head start the young man was tracked first to Haddington and then Edinburgh before sightings dried up and the trail went cold. A nationwide appeal was sent out with a description but not a single sighting of the rogue banker was reported, leading to the assumption he'd either killed himself or fled abroad. In fact, other than changing his name he was living quite openly in Gairloch, managing the local estate and selling steamer tickets to locals and visitors alike. That, my friends, is how remote Gairloch is.

If 'John Macleod' was around today there's a fair chance he'd have been collared by Alex Grey and cheerfully strongarmed into hosting a regular show on Two Lochs Radio. Officially Britain's smallest commercial radio station, Two Lochs serves an area that's similar in size to Glasgow but boasts a population of barely 1,600 people. The story of Two Lochs Radio is like a great Ealing comedy: a local community battling against impossible odds to realise a dream that should by rights have been impossible to attain.

I was the only person in the hotel the night I arrived. And I don't just mean the only guest: the woman who checked me in picked up her bag and keys and left immediately afterwards, returning the following morning to prepare breakfast. This was one of those marvellous old Scottish hotels, all wood panelling, animal heads on shields and paintings of rugged fellows in kilts standing amid rugged countryside beneath rugged skies and staring ruggedly into the middle distance. The stairs and floorboards groaned and creaked with every step as I headed out in the evening to a nearby bar for dinner, where a group of local men on stools at the bar were engaged in a detailed conversation about the shipping forecast. I walked back to the hotel warmed by whisky, giddily admiring the blanket of stars that seemed suspended just over the town.

By the time I woke the next morning the weather had closed in. In the bluish light of a Scottish winter morning the hills across the loch, silhouetted the previous evening by a glorious pink and purple sunset, had vanished entirely behind a murky curtain of low cloud. Rain spattered the window, propelled by a wind that made the old building creak.

I took out my little portable radio and turned the dial along the FM waveband. There was barely a hint of anything among the static as my thumb rolled the wheel until a crystal clear burst of powerful bagpipe-led rock all but sent me sprawling backwards over the bed. It had to be Two Lochs Radio. I knew from the listings online I'd consulted before I left for Scotland that Alex Grey, the station manager whom I'd be meeting a couple of hours hence, was hosting the breakfast show that day, so I set the radio down on the bedside table and sat by the window with a cup of tea, looking out at the *dreich* vista of a winter morning in Wester Ross listening to Britain's tiniest commercial station.

The voice that back-announced the song was exactly the sort that people have in mind when they describe a 'radio voice': calm, softly spoken and with just the right mix of authority and affability. It was also, considering we were in a remote part of north-western Scotland, surprisingly English. The news at the top of the hour detailed a new Gaelic language app seeking funding on Kickstarter, branched out into some advice on the dangers of cookies on your computer and the vulnerabilities of storing personal data on the internet, and gave notice of the local mother and baby group's next meeting as well as a retro disco and karaoke night taking place at a local hotel. There was Zumba with Janice, a gathering of the Step-It-Up Walkers and news of the impending arrival of a touring pantomime performed in Gaelic that told 'the traditional story of Little Red Riding Hood but with a martial arts twist'. The weather forecast followed, the nature of which reminded me, if I needed reminding, that I was in a pretty remote part of the country. First there was the inshore waters forecast ('wind northerly three or four backing westerly four or five, occasionally six later. Sea state moderate, weather fair with occasional rain, visibility good, occasionally moderate') and then the forecast for the mountains ('winds up to 40 miles per hour with little sunshine, poor visibility and little chance of a cloud-free morning').

This was a breakfast show unlike any I'd ever heard and I speak as someone who listened to Mark and Lard on Radio 1 during the nineties. The news may have been parochial – what else could it be when the total population of your catchment area wouldn't come close to filling the Albert Hall, even if they invited three friends each? – but it was slick, professional and in terms of production and presentation wouldn't have disgraced any number of national stations you could name.

The rain was falling steadily when I walked out of the hotel and made my way along the shoreline to the visit the Two Lochs studio. To say the station headquarters are understated is to understate an understatement. There's no two ways about it: it's a shed. A one-storey shed of whitewashed corrugated iron streaked with rust, a chimney at one end and planters either side of the front door.

'We have one brick wall,' Alex Grey told me, with resigned pride as he ushered me in out of the rain. 'The rest of it is literally just corrugated iron lined with hardboard.' He pushed open the door to his tiny office, realised the heater wasn't on then decided we'd talk in the studio instead – the only vaguely warm spot in the place. 'Officially this building is unfit for human habitation,' he said with a wry smile, edging round the desk to sit in the presenter's chair. 'But it's fine to use it for business purposes.'

Alex has been the station manager of Two Lochs Radio since it first went on air in 2003, having been one of the leading figures in establishing the station in the first place. A slim man in late middle age with an avuncular air, at first glance you could be forgiven for mistaking him for Richard Curtis. He's the only salaried employee of the station, which is otherwise staffed entirely by volunteers, and combines the day-to-day running of Two Lochs with the IT business he runs with his wife Anne, who is also a presenter on the station. In fact, he'd almost had to dash out during the breakfast show I'd listened to in order to attend to an IT emergency: 'I had an urgent call from round at the harbour where one of the businesses had suffered a complete loss of their internet service. As they in turn supply the internet to neighbouring businesses it was a bit of a situation: people couldn't take card payments or anything. In between presenting my links I was on the phone trying to get to the bottom of the problem then had to run round there and sort it out as soon as the programme finished.'

The studio had the same deadened sound as the ones I've used in Broadcasting House, yet here the walls were hardboard and covered in pictures and notices while the window directly behind the presenter's head wasn't even double-glazed. It was almost as if Two Lochs was pulling off the remarkable feat of continuing as a successful radio station through little more than graft, ingenuity, sheer will and probably a fair bit of wishing.

The Two Lochs story began during the nineties. As I'd found while tuning my radio that morning radio reception in this far-flung, low-slung part of Wester Ross is patchy at best, and even when they could pick up stations the locals found they were being ill-served even by ostensibly 'local' broadcasters.

'The original idea came from a man called David Carruthers, who now runs the convenience store down by the pier,' Alex told me. 'He had been the launch publicity manager for the Eden Court Theatre in Inverness, which is the cultural performance centre for the whole of the Highlands, and had done some work with Moray Firth Radio, the independent station that covers most of the Highlands. Then he came over here and was running one of the local hotels. Before long he'd decided that the circumstances particular to this part of Wester Ross meant there ought to be a station here, so he gathered together some likeminded volunteers and started kicking around the idea of launching a commercial radio station.'

By 1999 the idea had gained local traction, which is when Alex and his family moved to the area. It proved to be a case of the right man arriving at the right time. An electronics engineer by education, Alex's first job was introducing audiovisuals and working models to the Natural History Museum in London. 'Before that it was all stuffed animals and bits of card with writing on,' he said. 'I was brought in to help drag the place kicking and screaming into the twentieth century.'

It was there he met Anne, before joining the BBC at Alexandra Palace, producing working models and visual effects for the Open University, then relocating to Milton Keynes to become business systems manager for the BBC's Education Directorate. When the Corporation underwent one of its regular restructuring operations Alex had the opportunity to leave voluntarily on very good financial terms at a time when the family was considering moving to Scotland.

'My family's from Scotland,' he told me. 'My father's Glaswegian, and I'd always hankered after moving here one day. When the chance for voluntary redundancy arose one of our kids was just about to move from primary school to secondary, while the other was about to make the transition to A Levels, so we looked around the area and the school here was absolutely excellent. We moved to Gairloch in 1999. When I arrived the group trying the get the station off the ground said someone who knew the hot end of a soldering iron from the cold end was exactly what they needed.'

The group had managed to raise a fair bit of money but was struggling most of all with the sheer weight of bureaucracy involved in a commercial radio licence application. It was a challenge for a group of community volunteers and also new territory for the regulator Ofcom, or the Radio Authority as it was then.

'Persuading the Radio Authority to consider awarding a licence to the sort of project we were proposing was quite a hurdle,' recalled Alex. 'The Radio Authority itself was going out on a bit of a limb. There was no community radio back then and no such thing as a community radio licence. Something like the community licence they have today was being considered, they were looking into something called "access radio" at the time, but the only option available to us was a full commercial licence, the same licence held by stations like Capital Radio.'

This is where the story really branches off into Ealing film territory. Here you had a small group of volunteers clinging to one of the farthest fringes of the island of Great Britain, none of whom had run a radio station in their lives but who sought to establish one – the maximum listenership of which would fit comfortably into a barn. That station would have the same commercial status as a behemoth of an operation with a weekly audience larger than the entire population of Scotland. In terms of optimistic ambition this was a bit like your local minicab firm seeking to join the Formula One circuit with a couple of old Mondeos and a Citroën Picasso.

The Radio Authority didn't relax the process one iota, other than conducting interviews via telephone conference calls instead of making the Gairloch delegation travel to London and back every time. This was the only concession they were prepared to make to a group presenting a business model of a kind they had never seen before. For one thing, there was the premises: that a successful commercial radio station can be run from a little metal hut by the seashore makes 2LR nothing short of miraculous for that reason alone. I have sat in bus shelters bigger than the Two Lochs Radio building.

'Before we came along this building was a German language bookshop, for some strange reason,' Alex informed me just as it was dawning that nothing even tangentially concerning Two Lochs Radio could possibly surprise me ever again. A German language bookshop in a town of a few hundred people in the far north-west of Scotland?

Of course there was. 'The woman that ran it packed up and we took it on,' he continued. 'We ran it as a fundraising shop for two or three years, run by a couple of local ladies.'

I really hope they told the Radio Authority this. Used to reading business plans involving corporate sponsors, hedge funds and offshore investment conglomerates here they were faced with a station housed in a tin shed that until recently sold vernacular editions of Goethe and Schiller in which two women were now selling bric-a-brac and shortbread to passing locals.

'The application was a massive job,' said Alex. 'The pile of paper is an inch thick and runs to 150 questions, each of which has to be carefully researched and answered meticulously, because they're not going to award a licence to people they're not completely convinced can make a real go of it. With the multiple copies of the application you need to present to each board member in effect you end up publishing 20 copies of a bound book, and that's before they even consider you for an interview.'

All this, it turns out, was just to persuade the Authority to advertise a licence for the area, let alone award it to Two Lochs. The Two Lochs group faced another hurdle in that the area was technically already licensed to Moray Firth Radio, whose remit covered the whole of the Scottish Highlands. In order for the licence to be advertised, as part of their application the Gairloch group had to convince the Authority of Moray Firth's non-performance, that they were not serving their area. And even assuming the licence was advertised, there was the danger that a competing organisation might apply. Fortunately, that was never really going to happen.

'There were no other applicants for the licence,' Alex told me, 'which wasn't a surprise as there's no real commercial advantage for any companies that might have had even the slightest interest in broadcasting here, not even MFR, which is the only fully commercial Highland station.'

To the relief of all involved, after years of planning, campaigning and fundraising, the Radio Authority advertised the licence and granted it to Two Lochs.

That was the easy bit out of the way. Then the real work started, physical and financial.

'It was all done very much on a wing and prayer,' said Alex. 'The finances here to this day are extremely challenging because, although we're a tiny operation, the major costs don't scale down. We have the smallest Ofcom official population figure by far, but our licence fee doesn't reflect that. We pay the same for our licence as stations many times our size, and we pay the same level copyright charges on the music we play because the copyright societies when they set them never envisioned there would be radio stations as small as we are. Our annual turnover is about £40,000, but we're paying the same fees as a station with a turnover many times that figure.'

The peculiar circumstances of the station's catchment area also present challenges that would tax much bigger operations. A community radio station in Glasgow, say, serving the same number of people as 2LR would need a small transmitter, probably no more than five watts. In order to reach that same number of people Two Lochs needs four transmitters of two kilowatts each, requiring a large amount of electricity. Running a radio station, even one on this scale, is an expensive business.

'We rely a great deal on sponsorship from local companies,' Alex told me. 'It's hard because the sponsors we have don't really expect a commercial return, they're doing it more as a gesture of community support. Advertising is difficult for the same reason – advertising with us isn't really going to bring you new customers. Our local businesses don't need to advertise anyway, because there aren't two competing petrol stations, or butchers or whatever, so there's no real incentive there. We take regular advertising from the Scottish government, and we might get adverts from, say, Scottish Opera if they're touring in the area, but that's about it.'

2LR's income comes broadly from three sources: around a third of it is from advertising and sponsorship, another third from membership subscriptions and donations, while the rest comes from fundraising events.

'We have tombolas, bottle stalls and we held an autumn market this year, so it's a mix of things and a constant struggle,' said Alex. 'We get some heartwarming letters that often have money enclosed too, saying how much people appreciate us – but it's a shame, in a way, there's not a little coin slot on people's radios. We have a very well-established local newspaper which, of course, has a cover charge, but that's not a business model that will ever be open to us.'

Finances, although a constant headache, are only part of the challenge of starting and running a radio station. For Two Lochs Radio there were huge technical challenges to overcome, including having to place, install and maintain transmitter masts in order that the station can reach its listeners. 'The terrain here is absolutely horrible for radio signals,' lamented Alex.

The fact that he had broadcast both a mountain weather forecast and an inshore waters weather forecast on his breakfast show that morning provides a clue as to why the station faces its own peculiar logistical challenges. Most of the communities served by the station are coastal, huddled low down by the sea, but separated by mountains. This makes for incredible scenery but tricky radio.

'The towns and villages we serve are effectively separate entities as far as radio is concerned,' explained Alex. 'Almost all of them need separate masts on hilltops able to receive and send signals capable of reaching the audience. When we set up the station we had to do everything ourselves, from carrying the equipment up the hills to setting everything up and getting it running in a reliable fashion.'

Carrying radio masts up steep and rocky terrain before putting them together, raising them and making sure they're fixed sturdily enough to resist the wild weather that can afflict this part of the world was the easy bit. You then had to connect the mast to a power supply – power points being notably rare things at the top of mountains – and ensure the signal the transmitter was there to pass on was actually reaching the transmitter in the first place: in effect, making sure the masts could talk to each other. From the hut where we sat a microwave signal is sent to a mast on the hill across the loch that serves the Gairloch area. The signal is then bounced from there over the town and the hill behind it to a mast overlooking the town of Poolewe, a few miles to the north, from which Poolewe itself and the area around Loch Ewe receive their broadcasts. From there the signal is sent south to a mast down in Shieldaig that provides coverage to Loch Torridon and its surrounding area.

'Putting the mast up at Shieldag was a challenge. The transmitter cabin was in sections but was still very, very heavy. We'd had an offer from the local estate owner to borrow a quad-bike to get things as high up the hillside as we could before manhandling it up the last bit. The back wall of the cabin was two metres by two metres, this huge, heavy thing, and we had to go up the hill with four people behind the quad-bike holding it in place. The real Keystone Cops moment was laying the power cable up the hill. We had to dig a trench and Scottish Natural Heritage, who are understandably very keen to have as little damage to the landscape as possible, granted us one digger movement up and down the hill for the entire build. It went up to the top with everything laden on it, reels of cable, all sorts, before digging the trench on the way down. Of course, there was a thunderstorm and a few hairy moments when it looked like the digger would start sliding on the steep, wet surface, but luckily it just about kept its grip and the trench was dug.

'The next thing was to lay the cable in the trench and someone had the bright idea of starting at the top, so we'd be taking it down the hill rather than having to haul it up,' he continued. 'We mounted the drum on a pole at the top so it could turn and pay out the cable. A relay of people would then haul the cable itself down the hill, each person taking 50 yards before someone else would take it from them and heave it down another 50 yards. It's a thick cable though, very heavy, and once you had a hundred yards of cable on the ground you could hardly pull it at all. By the home straight we ended up with about five people all pulling on it at once because by that stage is was almost impossible to drag. Fifty yards from the end, just as we'd almost made it, one person stumbled, which stopped the cable, made everyone else stumble, and they all

225

dropped the cable. It took a superhuman effort to pull that cable the last few yards, I can tell you.'

All that was for just one of the three hill transmitters, and all long before anyone could think about playing records and giving out weather forecasts.

Finally, in 2003 after more than five years of planning, fundraising, and hauling equipment up mountains, the cables were laid, the masts lashed securely to the hills and the corrugated iron freshly whitewashed. Two Lochs Radio was ready to launch. Alex was one of the first voices to be heard bouncing between masts, over mountains and lochs and into people's kitchens and cars. He'd been involved as a youngster with one of the first hospital radio stations in the country, at Lewisham Hospital in south-east London, doing some presenting but keeping mainly to the technical side of the operation. He'd been inspired to become involved by the offshore pirate stations of the 1960s. 'I'd go to sleep listening to Johnnie Walker on Radio Caroline,' he recalled, and finally got behind the microphone, where he remains today well into his second decade as a radio presenter. He'd also come a long way from the BBC. And yet . . . not.

'Anne had been quite keen to get me away from the BBC because it just took over your life,' he said. 'Then we moved here and when we told people where we were living they said, "Oh, you're in the BBC house." It turned out that our house and the one next door had been built in 1964 by the BBC as the transmitter maintenance base for all the television transmitters on the west coast. Two houses, one for the senior engineer and one to be the maintenance base. The Director-General's signature is on our deeds. We still have a big pole in the garden with an antenna on top, and for the first few years Anne kept turning up great big glass valves whenever she dug the garden.'

For all the challenges in getting the station on air it's not all been plain sailing since for Two Lochs Radio. There are constant financial worries. They've rather outgrown the little shed, which is riddled with woodworm, and there's the constant pressure of keeping a full programme of presenters on the air.

'Money and human resources are the biggest issues,' mused Alex, 'and that puts understandable limitations on what you can expect from people even once you've persuaded them to get involved. It's quite a commitment; presenting a radio programme as part of a schedule is not the kind of volunteering that invites flexibility, especially something like a breakfast show which involves antisocial hours, and while our presenters do work very well at trying to fill in for each other when people have other commitments it can be tricky.'

Two Lochs draws its presenter base from a population of less than 2,000 people, not all of whom are a stone's throw away from the studio. Alex estimates that at least a hundred presenters have passed through since the station opened, which means that one in every 16 or 17 people you pass in the region is, or has been, a radio presenter.

'I can only remember one or two that really weren't cut out for it,' said Alex, 'but it's very rare. It's more often the opposite problem, someone you know would be great but they look at their shoes and mumble, "Oh no, sorry, that's not for me." We're having a bit of a drive right now as, without more people, we'll reach a bit of a crunch point fairly soon. I've had three people put their heads above the parapet in the last couple of weeks, so hopefully at least one of those will come through.

'The breakfast programme is the difficult one to fill because it means being up at six in the morning when it's pitch dark outside. Sadly the person who's done two breakfast shows a week for us for the last ten years is about to move away from the area. The breakfast show is a bit faster than the others too,' he explained. 'It works on a half-hourly cycle with the news, the inshore forecast, the mountain weather, local events and so on, and maybe one music track played between some of those things. If you haven't prepared properly it can be a bit hectic. The drivetime programme between five and seven, however, is a lot more relaxed and people seem to prefer that.'

Again, we're in the topsy-turvy world of Two Lochs Radio. Just about every other station in the world will have a queue of people round the block desperate to have a crack at the breakfast show, to have their photograph taken wearing pyjamas while yawning theatrically with a giant comedy alarm clock, jolting the listening audience out their nocturnal stupor with some slick sounds and high quality bantz. I was a little amazed to find that Two Lochs often struggles for presenters. I was on the point of volunteering for a couple of breakfast slots myself, upping sticks and heading north, all the way from Kent with some solid gold banterbury tales, maybe even doing a John Macleod and ripping off a bank on the way. I was just about to open my mouth to offer this to Alex – I'd even bring my own alarm clock for the photos – when sense prevailed. For one thing, my wife would probably start to wonder where I was. But surely, I asked, with the number of disc-jockeys and radio presenters kicking their heels around the country, not to mention the armies of graduates being churned out by the universities every year, he must be swamped with CVs and presenter demos?

'There was a time when we did get quite a bit of that, people wanting to come here for work experience,' he said. 'But the problem is we don't have the resources to support someone in that situation. Similarly, we've had plenty of school-age pupils coming through the stations and some of them have been really good, but by and large they leave the area when they go off to university and don't come back – or when they do they have other things to do here and can't commit to regular presenting.

'We used to have a regular Saturday morning crowd of young presenters – some of their scripts are still pinned up on the wall here – and they were fantastic. Some have gone on to do media studies at university, one's working in film and was recently nominated for a BAFTA, another's gone into sound engineering, and for many of them the community service aspect goes towards their Duke of Edinburgh awards – so it has given some of the local kids a good start to their careers, a bit of extra embellishment for their CVs.'

It's even harder to understand the concept of a commercial radio station that sometimes struggles for presenters when you consider that Two Lochs doesn't broadcast its own material around the clock: overnight and on weekdays between the breakfast show and the drivetime show the station takes the feed from Smooth Scotland. With the station's own programming, however, the key for Alex is variety.

'The underlying principle here is to broadcast programmes you're not going to find on the BBC,' he said. 'There's no other local radio here, no commercial radio, no BBC local station, so whatever we put out, if it's something people won't get on one of the existing stations, we're fulfilling the audience's requirements. Maintaining choice is a big principle, and that goes for Ofcom too – they don't want two stations broadcasting pretty much the same thing as that wouldn't be good from the point of view of the community. From our point of view it would also be a waste of a licence if we were just churning out stuff you could hear anywhere.

'Our demographic is completely wide open. Maybe, on balance, it veers towards the older end of the spectrum, but it is very broad and over the years our presenters have ranged from eight to 80. One of our regular presenters at the moment is actually over 80. We also provide a home for people who've been pensioned off by major commercial stations, such as those run by Saga. The one in Glasgow which is now Smooth Scotland, the station we use as a sustaining service when we're not on the air during the day and overnight, had some very good themed programmes – book reviews, rock 'n' roll, and the country music presenter Bill Black had a very popular programme before Saga in their wisdom decided he didn't fit their profile any more. Now he produces his programme independently and we're one of the stations that carries his show. We have a Saturday rock 'n' roll show with a guy called Mike Marwick that comes live from his attic over in Kinross: by day he's a property solicitor, by night a raving rocker. We have a guy in Florida who does a country–roots programme, and a pair in Cape Breton who do a weekly Gaelic programme. We have a local Gaelic show too; generally it's older folk who come in and chat about old times, memories of schooldays, local traditions, that kind of thing. I think we're up to edition 580 of that programme now and it shows no sign of stopping anytime soon.'

It's this hyper-local aspect that's the key to the success of Two Lochs Radio. With a small population that, while concentrated in relatively large communities, is spread fairly widely over a rugged landscape, a programme of Gaelic language reminiscence seems of vital importance. A pensioner living on their own, possibly in an out of the way part of the region, can tune in to a programme like that and feel the intimacy of their community, immerse themselves in their shared personal stories, without leaving their armchair. The familiar voices, the familiar memories, it's a vital service they're not going to find via Radio 2, Classic FM or Radio Scotland. The Two Lochs listener is one who likes to widen their horizons while at the same time being perfectly happy with the horizon they can see out of the window. That's why, for all its challenges, the station is a success. I think it's also why the people involved, in particular Alex himself, work so hard to keep it going: they know what the community would lose if the station ever gave up the ghost. It's why, even when something goes disastrously wrong, the community makes sure that, no matter how heavily stacked the odds might be, the show goes on.

In January 2005, two years after the station launched, it suffered what could have been a cataclysmic blow when hurricane-force winds flattened a transmitter cabin up on one of the hillsides. Power to the whole region was knocked out for five days, leaving everyone facing freezing temperatures at a time of year when there is no more than five hours of daylight. When the power finally returned everyone at Two Lochs was delighted to find that the transmitters, which had been right in the teeth of the storm, came back on with it.

'We thought, "Wow, that's amazing, but we'd better go and check it out." I went up one hill with my wife and daughter, and our daughter strode out ahead, stopped, turned round and called out, "It's not looking good, Dad." I wondered how she could tell, because from where she was standing she wouldn't have been able to see where the cabin was. When we caught up we saw that she was standing over the door, bent, buckled and laying on the ground a good distance from where it should have been. When we reached the top the whole thing had gone over into a parallelogram – only the equipment rack inside had stopped the building from being completely pancaked. Despite being covered in mud and rain and everything the elements had thrown at it, the rack had burst back into life when the power came back on, lying there in the mud. I knew it was hopeless, though: its fans had sucked so much mud and water through that it had corroded and become a complete write-off. We had to start all over again and put up a brand new, bigger, stronger and heavier transmitter cabin. This one wouldn't go on the back of a borrowed quad-bike, we'd need a helicopter, and that would cost an absolute fortune, probably more than we could afford at the time.

'It was David Carruthers, the man who'd had the idea for the station in the first place, who saved the day. Helicopters are used quite regularly up here to transport building materials to the top of hills, so we knew it was feasible, we just needed to get the cost down as much as we could. A mixture of charm, persuasion and a carefully expressed fax meant that, in the end, we got the helicopter for a cost that we could just about afford, as long as we were on the hill at the crack of dawn on the appointed date. We only had a 20-minute slot: the helicopter had to arrive, pick up the cabin from the bottom of the hill, deliver it to the top and be away again to its next job all within the space of 20 minutes.

'So there we were, before dawn, in the dark and the freezing cold, standing on top of the hill while the cabin stood at the bottom waiting for the helicopter. Then, far in the distance, somewhere over Poolewe, we saw the lights of the helicopter rise above the hills and become gradually larger as it started to come towards us. We heard the thump of the rotors over the wind and it was just like *Apocalypse Now*. Someone even started singing Wagner.

'The cabin was hooked onto the chopper and up it came. It needed to be positioned close to the mast, no more than 15 metres from the guywires holding the 20-metre mast in place, and because once it was on the ground we had no way of moving it the chopper pilot had to get it spot on, in the dark, inside 20 minutes. He was brilliant. He flew round at an angle so the cabin swung out, like a fairground seat on a chain, coming in on a curve and dropping it right on the spot.'

There's a deeply impressive spirit at Two Lochs Radio that affirms your faith in the medium and its potential. There's a vision and determination here of the kind that led to the offshore pirate stations of the sixties, or that period in the eighties when nearly every tower block in every city had a transmitter on the roof with people playing records through it from one of the flats. When radio isn't serving a particular region or demographic in the way it should, there are people out there quite prepared to go off and make their own radio instead. Some go the pirate route, others, like 2LR, choose the arguably more difficult route for a small concern of taking on the bureaucratic process and emerging at the other end, exhausted, bleary and displaying symptoms of mild shellshock, but waving a licence. A decade and a half on and Two Lochs has faced down just about every challenge a small radio station run on a turnover adding up to less than two people living on the UK average wage could possibly imagine and is still there, broadcasting its mountain weather forecasts and country songs and Gaelic reminiscences to Britain's smallest radio audience. It's come through a crippling national recession and hurricane-force storms and emerged relatively unscathed. Despite this, and despite the happy-go-lucky Ealing angle I've superimposed on the story, Alex Gray is as much of a realist as he is an enthusiast. The next challenge is always just around the corner and he is under no illusions.

'A lot depends on the next few months,' he told me after a tight-lipped pause when I asked him about the future. 'We've lost a regular sponsor recently, one who had provided us with sponsorship from the very beginning, which is a big blow. We have a person moving away and another one who we thought would become a regular presenter has decided it's not for them after all. Right now we're on a low both in terms of income and a team of regular presenters, and there'll be some crunch meetings in the next while.

'There are a few encouraging signs, mind you, not so much on the money front, where we might have to cut a few more costs where we can, but we've had a few expressions of interest on the presenter side so there's a good chance we'll get back on track.'

The commitment shown by all involved with Two Lochs, from Alex combining running a commercial station with his own entirely separate IT business, to the presenter coming in for an hour a week, every week, with all the inconvenience that iron horse in their weekly diary can sometimes bring with it, this is radio as a community service. It's a commercial station according to the letter of its licence, but commerce here extends little further than a local business sponsor pushing a bit of spare cash the station's way in the full knowledge that it's not going to prompt hordes of new customers to come hammering on their door.

Alex Gray is the perfect station manager, especially for a concern like Two Lochs. He's smart, practical, empathetic, persuasive and realistic. Buried under all that, but not too deeply, is the boy listening to Radio Caroline through a transistor under his pillow.

'I'd go to sleep listening to Radio Caroline and wake up to the Home Service,' he recalled with a smile. 'I'd also listen to *Listen with Mother* when I'd visit my grandmother in Lancashire. She ran a shoe shop, and shops back then would close for lunch for an hour at 1 o'clock and *Listen with Mother* was on at 1.45, so we'd listen to it together before she went back into the shop.

'Believe it or not, my mother used to listen to *Coronation Street* on the record player,' he continued. 'When I was very young we lived in a flat on Beulah Hill in south London in the shadow of the ITV transmitter. We didn't have a television back then but the signal from the transmitter was so strong that thanks to dirty connections picking up the transmissions the sound would come through the record player. This meant my mother was well-versed in the goings-on in Weatherfield before we even owned a television.'

Following *Coronation Street* on a record player? That sounds almost as nuts as starting a commercial radio station for a remote area that could almost have been topographically designed specifically to prevent such a thing, aimed at a total potential audience roughly equivalent to the one that watches Accrington Stanley on a Saturday afternoon.

19

The Birth of Satire: Ronald Knox and the Red Panic of 1926

One of the stories that is always trotted out whenever people want to talk about the power of radio is the reaction of the public to the broadcast Orson Welles's adaptation of *The War of the Worlds* on CBS on 30 October 1938. America, the legend relates, went nuts. There was utter bedlam. So convincing was Welles's depiction of alien invasion it had people running out of their houses and into the streets, thousands of them in every city, running into each other like Laurel and Hardy when the rent collector knocks at the door, screaming and rending their clothes, convinced there were Martians swarming over the country zapping people at random with giant ray guns. So powerfully rendered was Welles's dramatisation there was panic across the US on a scale never previously seen; a national nervous breakdown in all but name.

'Radio Play Terrifies Nation' screamed the front page of the *Boston Daily Globe*. 'Fake Radio "War" Stirs Terror Through US' bellowed the *New York Daily News*. The *Amarillo Globe Times* went for the wordier 'Thousands Flee Homes, Pray or Faint as Fictitious Radio Program Relates Invasion of Martian Hordes', while the *Akron Beacon Journal* was so spooked it splashed 'Radio Paints Livid Picture of Attack, Panic Follows' when they probably meant 'vivid'.

Except, well, it didn't really happen that way. There wasn't any mass panic. There wasn't much in the way of terror, either. A few people might have looked askance at each other across their living rooms, some might have gone to the window and looked up at the sky, others possibly phoned a neighbour to see if they'd heard this thing on the radio, but reports of panic, screaming, cars being overturned, hat brims chewed by wild-eyed citizens, whiskery men in plaid shirts and long johns blasting at the stars with blunderbusses . . . they were all grossly exaggerated. According to a survey that night by the radio ratings company C.E. Hooper, only two per cent of the radio audience was listening to *Mercury Theatre on the Air*. Nearly everyone else was listening to NBC where ventriloquist Edgar Bergen and his dummy Charlie McCarthy were the sensation of the day on *The Chase and Sanborn Hour*.

Now, any nation that tunes in *en masse* to hear a ventriloquist on the radio might seem like a gullible bunch of hayseeds ripe for hoaxing, but the sheer number of radio ventriloquism devotees meant that the Martian tripods spidering across the country went almost completely unnoticed other than by earnest drama fans listening with pursed lips and steepled fingers, too cynical by half to be perturbed by something as vulgar as an alien invasion.

In truth the cities of the US that night were largely no busier than usual for an American Sunday evening in autumn during the late thirties. No switchboards were jammed by panicking callers, hospitals were not overwhelmed by hordes of pale people clutching at their chests or claiming laser burns, and the National Guard didn't take to the streets in an effort to restore order. Nobody topped themselves at the prospect of subjugation to Martian rule. It was all, in the main, a load of cobblers.

Why has the myth propagated and persisted for all these years? Essentially, while the story was a non-starter the newspaper headlines were very real, and it all came down to those two key drivers of everyday life: money and fear. Specifically, in this case, the fear of losing money. The phoney-baloney stories of wild-eyed crowds careering through the streets of towns and cities across the country were the newspapers grabbing an opportunity to swing a hobnailed boot at the medium of radio, something that had grown massively in America during the 1930s, turning the heads of the public and, most importantly, advertisers away from traditional printed news sources. If radio could be shown to have caused widespread panic across the nation by irresponsibly hoaxing the public into thinking extraterrestrials were stalking the land taking laser pot shots at grain stores, water towers and Cletus's moonshine still, well, maybe those same people would go running back to those same reliable, trustworthy newspapers again for reliable and trustworthy news.

The earnest accounts of mass panic may have been largely hokum but they made for an undeniably good story. Nineteen thirty-eight was one of those years in which everyone was jumpy: international politics were unstable to the east and the west, people's default setting was 'edgy' and here was Orson Welles creeping up behind them and bursting a paper bag next to their earhole. It's no wonder the nation's newspapers glared through their lorgnettes with haughty indignation at the upstart actor and producer: suddenly the moral high ground was not only there to be occupied, it came with a competitive rate card for advertisers.

With only the documentary evidence of newspapers to go on, the Martian panic myth has over the years solidified from the opportunistic exaggeration of a flimsy premise to hard fact. Indeed the *War of the Worlds* story has become such a totem of broadcasting anecdote it's completely eclipsed the *original* radio panic, one that predated it by 12 years, which was broadcast by the BBC and emanated from the unlikely figure of a genial classics scholar turned man of the cloth.

As 1926 dawned there was a definite sense that something was afoot in Britain. Industrial unrest had been on the rise since the end of the First World War and, like most countries in Europe after the Russian Revolution of 1917, the establishment was looking nervously over its shoulder fearing the Bolsheviks were creeping up behind them with a length of lead piping. The First World War had seen officers and men living in the trenches muddy cheek by grimy jowl, but when the men returned home to take up their preordained places in the rigid class structure of Edwardian Britain the almost feudal sense of deference that kept the class barriers firmly shored up had begun to erode. The Labour Party was getting stronger. The Communist Party of Great Britain had been founded in 1920. In 1924 the *Daily Mail* had published a letter apparently from the head of the Comintern in Moscow, Grigory Zinoviev, to the CPGB urging it to undertake seditious activities and encourage the radicalisation of the working class ahead of the general election that year. The letter was a forgery, but at the time it was enough to stoke establishment anxiety of a popular uprising and a general fear of the red hordes.

The previous year the government had attempted to ease some of its economic problems by joining the gold standard – a move that backfired, triggering economic depression and a decline in exports, particularly of coal. In the summer of 1925 the coal industry moved to cut wages and increase working hours, and a General Strike in support of the miners seemed on the cards until the government stepped in to prop up wages in the industry.

There was a distinct sense of nervousness among the middle and ruling classes at the dawn of 1926. They could sense something simmering, a threat to what they perceived as the natural order of things. For the first time the working classes had the facility to organise: established social order seemed suddenly flimsy.

On 16 January at 7.40 in the evening, Father Ronald Knox, a Catholic priest and an eloquent, witty broadcaster, sat down in an Edinburgh studio to give a talk called *Broadcasting the Barricades*. Unlike most radio talks of the era this wouldn't simply be a man droning drearily into the microphone. This one would go further, even featuring rudimentary yet groundbreaking special effects. Looking back, it's arguably even the first scripted radio comedy, certainly the first radio satire, containing as it does specially written characters and events rather than a couple of music hall veterans with cheesy catchphrases tossing terrible puns back and forth across the studio.

The broadcast began with a dreary, lisping male voice emerging from static holding forth lugubriously on the subject of Gray's 'Elegy . . .' At the conclusion an announcer's voice thanked Mr William Donkinson for his lecture on eighteenth-century literature, declared that he would be continuing with the news, gave a Test match score in which England were having their backsides handed to them by Australia, then related a story of a young girl being rescued from the Thames at Chiswick. So far, so plausible, other than the fact that the England cricket team wasn't in Australia during the winter of 1925–26.

Then things took a turn for the alarming.

'The unemployed demonstration,' intoned the announcer, 'is now assuming threatening dimensions.'

He went on to describe a mob of unemployed men gathered in Trafalgar Square marshalled by a Mr Popplebury of the National Movement for Abolishing Theatre Queues that was, as he spoke, being whipped up into enough of a fervour to attack the National Gallery. He gave a brief description of the history of the gallery before adding, 'It is now being sacked by the crowd on the advice of Mr Popplebury, Secretary of the National Movement for Abolishing Theatre Queues. That concludes the bulletin for the moment, you will now be connected with the band at the Savoy Hotel.' At which point dance music emerged from the nation's radio sets.

A weather forecast followed, predicting hurricane-strength winds right across England and Scotland, before an update on 'the unemployed demonstration' told of a crowd surging through Admiralty Arch then spilling into St James's Park where it threw bottles at the ducks. 'So far no casualties have been reported.'

The announcer then went on to trail a lecture by Sir Theophilus Gooch, Chairman of the Committee for the Inspection of Insanitary Dwellings, on the subject of 'housing for the poor'.

'Eh, what's that?' the announcer said. 'One moment please . . . From reports which have just come to hand it appears that Sir Theophilus Gooch, who was on his way to this station, was intercepted by the remnants of the crowd still collected in Trafalgar Square, and is being roasted alive.'

A potted biography of Sir Theophilus Gooch followed – 'he very soon attracted the notice of his employers, however nothing was proved and Sir Theophilus retired with a considerable fortune' – at the conclusion of which came confirmation he was being roasted alive in Trafalgar Square 'and will therefore be unable to deliver his lecture to you on the housing of the poor. You will be connected instead with the Savoy Band'. A news update followed the music, an American film star arriving by liner at Southampton after a 'capital crossing' and then an update that the crowd under the direction of Mr Popplebury was now destroying the Houses of Parliament with trench mortars. The Clock Tower had fallen, Big Ben was destroyed. 'Greenwich time will not be given from Big Ben this evening,' he continued, 'but will be given instead from Uncle Leslie's repeating watch.' (Uncle Leslie was a popular children's presenter on the Edinburgh station.) Another update came in: Mr Wotherspoon, the Minister for Transport, had been caught by the mob trying to flee in disguise and was consequently hanged from a lamppost on Vauxhall Bridge Road.

'One moment please . . .' There was a pause. 'The British Broadcasting Company regrets to announce that one item in the news has been inaccurately given and the correction now follows. It was stated in the bulletin that the Minister for Transport had been hanged from a lamppost on the Vauxhall Bridge Road. Subsequent and more accurate reports show that it was not a lamppost but a tramway post which was used for this purpose.'

Then it was back to the Savoy Band, whose performance ceased suddenly with a loud bang.

'Hallo everybody!' said the announcer. 'London calling. The Savoy Hotel has now been blown up by the crowd at the instigation of Mr Popplebury of the National Movement for the Abolition of Theatre Queues.' He then related that the more unruly elements of the crowd were approaching the BBC headquarters at Savoy Hill. 'One moment please,' he intoned. 'Mr Popplebury, Secretary of the National Movement for the Abolition of Theatre Queues, with several other members of the crowd, is now in the waiting room. They are reading copies of the Radio Times. Goodnight everybody, goodnight.'

Looking back nearly a century later it's clear to see the roots of some of our finest broadcast satire here: Chris Morris with On the

Hour, *The Day Today* and *Brass Eye*; there are even traces of the Goons and *Monty Python*. It's a brilliant piece of work: enough 'reality' to keep things plausible, enough nonsense for people to know that it's satire.

And yet . . .

Some people took the broadcast at face value. Despite a rampaging mob being led by a man called Popplebury from an organisation opposed to queuing outside theatres, despite the supposed immolation of a man that didn't exist from a committee that didn't exist, despite the lynching of an entirely fictitious Minister for Transport, despite the recondite and earnest correction of his fatal suspension being from a tram post rather than a lamppost, and despite the broadcast ending with Mr Popplebury waiting patiently outside reading the *Radio Times*, some people actually took the broadcast at face value.

While there were no accounts of actual panic in the streets, the newspapers reported a clutch of incidents in the following days. The Mayor of Newcastle, for example, returned home from a dinner to find the Lady Mayoress 'greatly upset' by the broadcast. One aggrieved correspondent noted that only recently had he presented some elderly relatives with a radio set, and Father Knox's broadcast was only the third time they'd used it. 'It is in the lap of the gods,' he sighed, 'whether their health is not permanently affected by the shock.'

The BBC received a flurry of calls, as did the newspapers, while the Savoy Hotel later revealed that it fielded more than 200 phone calls enquiring after its fate, from concerned relatives of people staying at the hotel to those worrying whether future bookings would be honoured.

The BBC issued an official statement. While it flat-batted the situation in a suitably apologetic tone, pointing out that the broadcast was preceded by a notice that the ensuing item was a skit, it was clear the fledgling company was sighing with exasperation that members of the Great British Public had actually been taken in. The Corporation concluded with the immensely comforting, 'London is safe, Big Ben is chiming and all is well,' which, if you ask me, in the tumultuous times in which we live today is a much more reassuring slogan for mugs, tea towels and aprons than 'Keep Calm and Carry On'.

An anonymous BBC spokesman quoted in the newspapers during the aftermath went a bit further than the official line, telling the reporter with an almost audible eye roll, 'I feel the sense of humour of the British people will stand an even more serious attack than this. We take the view that of the roughly two million people listening in, quite nine-tenths saw the humour of the thing. They, of course, remained silent; it was the one-tenth who missed the joke who raised the scare.' Another missive from Savoy Hill stated that the company had received a great deal of correspondence in the aftermath of the broadcast, but 'the letters of appreciation outweigh the number of those written in protest', and another noted that, 'Most of the letters of protest received by the BBC come from England. A few come from Wales but, generally speaking, Scotsmen do not seem to have been misled into making a catastrophe of the burlesque.'

Rumours continued to abound: Dublin's milkmen were apparently relating the tumultuous events in London on doorsteps across the Fair City up to a week after the broadcast; others because bad weather the following day meant that many areas didn't receive their Sunday newspapers – in the circumstances a clear consequence of the insurrection in the nation's capital that had reduced the Houses of Parliament to rubble.

The man behind the broadcast, the creator, writer and the provider of all the voices, affected surprise that anyone saw what the newspapers called 'a burlesque that was a skit on broadcasting' as anything other than satire. 'I should have thought that not even a child could be taken in,' said Father Ronald Knox.

Knox was a remarkable man; the fact that the first biography written of him was by Evelyn Waugh is an indication of that. A celebrated classicist, noted satirist and a writer of detective novels so highly respected that his 'ten commandments' of crime fiction are still famous among thriller writers today, Knox was a quite brilliant polymath. Educated at Eton and at Oxford, his skill in the classics was so renowned that he was engaged as a tutor to a young Harold Macmillan. An enduring friendship was forged that meant many years later when Macmillan was Prime Minister and Knox was dying of cancer he put up his former tutor at 10 Downing Street while Knox visited London specialists. The son of the Anglican Bishop of Manchester, Knox was ordained into the Anglican church in 1912 but converted to Catholicism in 1917 following intense and lengthy theological discussions with his friend G.K. Chesterton. The decision prompted his father to immediately cut him out of his will, an act that didn't deter Knox from being ordained as a Catholic priest in 1918.

Not only did he translate the Bible into modern English – 'a sort of timeless English that would reproduce the idiom of our day without its neologisms' – he displayed his innate gift for satire by writing critiques of the Anglican church in the styles of Dryden and Swift. Annoyed by the emergence of those questioning the authorship of Shakespeare's plays, Knox published an earnest and, on the face of it, convincing treatise that proved via cryptology that Tennyson's 'In Memoriam' had, in fact, been written by Queen Victoria.

In the week that followed the broadcast Knox had already arranged to give a number of speeches around the country on the subject of satire and its place in British society. 'An Englishman laughs at the remarks of an Irishman that the Irish do not recognise as humorous,' he told an audience at Sheffield University. 'Consequently there is a tradition that the Irish are incorrigible humourists incapable of governing themselves. Scots, on the other hand, have an unfortunate habit of ruling Englishmen, so are considered to have no sense of humour.'

The British press were less sensationalistic than their American counterparts but their condemnation was as enthusiastic as it was motivated by the same market forces. 'The British Broadcasting Company has lent itself to a practical joke of a particularly foolish character,' chuntered the *Daily Express*. 'Why it should seem funny to a priest to insult the working people of England by representing them in the act of dynamite outrages is one of those problems we do not pretend the ability to solve.' The *Daily Sketch* noted how the incident proved the 'Briton's faith in his press'. 'In future,' it continued smugly, 'they will know better than to assume anything is "news" until they see it in print.'

Across the Atlantic the *New York Times* noted that it was a good story that the British newspapers seized upon with glee. 'It was thoroughly representative of this stodgy old country', huffed the paper's London correspondent, 'and also the dear old ladies who have abandoned knitting needles for wireless sets.'

Radio was in its infancy in the early days of 1926 and news radio was barely out of the womb. The broadcast of news over the airwaves, particularly news as it was apparently happening live, was an unprecedented concept for the BBC audience. Add to that the use of satire being completely unknown on radio at the time, and it is little wonder that some of the listening audience was convinced that London had fallen under the yoke of the Red Menace.

Ronald Knox had pitched it exactly right. The class and industrial tensions binding the nation were pulled taut enough to be twanged by a skit like the one he devised. The newness of radio made such a broadcast plausible to listeners, even when the sound effects were little more than a man smashing up a crate and dropping a cloth bag of broken glass to represent one of the world's leading capital cities falling to a riotous mob.

So good was Knox's broadcast that it deserves to be better known than its *War of the Worlds* counterpart that followed a dozen years later (the *Nottingham Evening Post*, incidentally, with astounding prescience, described the Knox broadcast as being 'like a page from a Wellesian Martian romance'). It was better executed, more plausible and was heard by the entire radio audience of the day. However, it also deserves to be known as a groundbreaking piece of broadcast comedy satire. *Broadcasting the Barricades* was decades ahead of its time.

20

'The Goons are the Lonnie Donegan of British Comedy': Arthur Mathews in the Comic Ether

A warm Saturday teatime tinged with hints of approaching autumn: long shadows, golden light and a calm sea of the deepest blue. It was the farthest end of summer at the farthest end of the promenade at the farthest end of Bexhill-on-Sea, and I was slithering around in canvas shoes atop a muddy, gorse-covered hummock yards from the cliff edge. I cut a curious figure, prowling back and forth staring into a dense clump of trees and bushes, occasionally bending over, hands on knees, screwing up my eyes and peering into the undergrowth. At best I looked like a man who'd lost his dog; at worst that I'd heard that this was a place where a gentleman might come to find a particular kind of no-strings companionship. The sideways looks I was getting from the skateboarding kids as they pendulumed back and forth in the nearby half-pipe certainly suggested the latter.

My aims involved neither the canine nor the supine, however: I was looking for an old concrete structure bedded long ago into the top of Galley Hill, outwardly charmless and, in the one photo I'd seen, covered in graffiti, with its doorway and windows filled with breezeblocks.

What I sought was one of the hundreds of Second World War pillboxes that lined the vulnerable coasts of Britain, strung along the southern edge of the country so that no part of the sea could be left unobserved by nervous conscripts scanning the horizon for a German invasion that would never come. Even the most ardent enthusiast for these buildings would struggle to make a case for their pulchritude, and certainly my own interest in Second World War architecture veers somewhere between minimal and non-existent.

This one was different, though. It wasn't so much the building itself that had brought me to this muddy spot, it was more the person who had once occupied it many years before. In two years stationed at Bexhill-on-Sea during the war he'd spent many long and boring hours here, relieving the tedium by writing jokes, poems and sketches – the kind of thing that would go on to make him a national treasure and lead to some of the greatest and certainly most influential radio comedy in the history of the medium.

The pillbox that was evading me somewhere in the gorse was where Spike Milligan had spent long periods of his early military service during the Second World War. He'd written about it in the first volume of his war memoirs *Adolf Hitler: My Part in His Downfall*, one of the earliest books I remember pulling from my parents' shelves. I loved it, reading and re-reading the book until it fell apart, even though I could have been no more than ten years old and didn't get most of the jokes. For one thing, when he returned from his military childhood in India Spike's family settled in Catford, a part of south-east London just round the corner from where I grew up. The book was littered with references local to me: New Cross, Deptford, even Chiesmans department store in Lewisham round which I'd often be dragged reluctantly by my mother as she shopped for bedlinen or curtains.

A couple of years or so later I also acquired an LP called *The Laughing Stock Of The BBC*, containing snippets of old BBC radio and television comedy ranging from *Round the Horne* to *Not the Nine O' Clock News* via *Hancock's Half Hour* and one of Noel Edmonds' prank phone calls from the seventies in his days at Radio 1.

The album also contained two extracts from *The Goon Show*: anarchic scenes from episodes called 'Napoleon's Piano' and 'The Case of the Missing CD Plates'. I played the record until the needle wore out and before long I could recite the whole thing perfectly. The clips that really stuck in my mind were the *Goon Show* extracts.

With the surreal anarchy, the near hysteria, the bizarre voices and the playfulness with the concept of radio to conjure bizarre images in my imagination, it was like nothing I'd ever heard. It was punk rock, except the band members had scripts in their hands instead of guitars.

It was only later that I learned Spike Milligan was the man most responsible for this epoch-subverting freeform humour that was in such contrast to the more conventional tone of the rest of the record and the comedy I was watching in the early eighties. Television back then was fairly staid, and while I'd sit dutifully through *Terry and June*, *To the Manor Born* and *Last of the Summer Wine* I didn't find them remotely amusing. The fallout from the similarly revolutionary *The Young Ones* had yet to settle and spread but these incredibly inventive, freeform snippets of Milligan's mind changed the way I saw the world for ever. *The Goon Show* whipped off the tablecloth of the world, sending conventionality crashing to the floor and leaving behind all its absurdities.

Hence, when I commenced the search for the true birthplace of radio comedy I'd ended up here, at an apparently unremarkable spot on the south coast, watching the shadows creep across the grass as the sun dipped towards Beachy Head further to the west. Somewhere among the thickets in front of me was the place where Spike's mind had wandered to help pass the tedious hours of staring at the horizon and watching the same shadows making the same diurnal progress that now included my own. Here, on the outskirts of Bexhill-on-Sea on a hillock between the sea and a retail park was the place where *The Goon Show* and all the extraordinary radio comedy that followed it was born. Milligan even set one of the episodes here, 'The Dreaded Batter Pudding Hurler of Bexhill-on-Sea'.

As both darkness and temperature began to fall and the lights on the corrugated grey superstores on the other side of the hill began to sharpen and brighten in the gathering gloom I gave up the search. I'd heard a rumour that the pillbox had been buried by the gas board a few years earlier. Maybe that was true, after all, and the observation post was now somewhere beneath my feet. Instead, I sat on top of the hill and looked out at the English Channel, taking in a view that wouldn't have changed since Milligan was stationed here as a member of D Battery, 56th Heavy Regiment, Royal Artillery.

According to his memoir, which I pulled out of my bag, trying not to lose any of the loose pages in the breeze, Milligan was posted here just as the Dunkirk evacuation began. He remembered a warm, still day, one on which he and his colleague went for a swim in the sea, the evacuation invisible over the horizon but the low booms of explosions carried to them across the water on the breeze.

It was here in Bexhill that Terence Milligan acquired the nickname 'Spike'. Back at the regiment's barracks – a requisitioned girls' school in whose library Milligan would devour Dickens and George Eliot – he and fellow members of the jazz band he'd started would tune in to Radio Luxembourg and listen to Spike Jones and his City Slickers. Somehow the name was attached to the Bexhill battery band, and Spike Milligan was born. Also during his two years based in Bexhill Spike sometimes passed downtime in the barracks and at the observation posts reading 'Popeye' comic strips, which is where he came across a big, dopey babysitter character named Alice the Goon. The name lodged in his mind and would appear frequently in his notebooks. The word 'goon' would become one of the most famous in the land, and certainly the most influential in British comedy – there are traces of *The Goon Show* everywhere in broadcast comedy today.

There had been radio comedy before the Goons, of course, but Spike Milligan, Peter Sellers, Harry Secombe and, in the early days, Michael Bentine were the first to recognise and harness the true *magic* of radio to make people laugh. From the start Milligan utilised sound effects to help create the world of the absurd inhabited by the likes of Neddie Seagoon, Major Bloodnok, Bluebottle and Eccles. The scripts were full of incredibly inventive characters, word play and scenarios that brimmed with genuine nonsense. It takes a rare talent to render nonsense actually funny and Spike Milligan was a master.

When the first *Goon Show* went out (as *Crazy People*, a title insisted upon by a BBC sceptical that anyone would tune in to a show with a name that didn't remotely mean anything) on the Home Service at 6.45 p.m. on 28 May 1951 – slotted between a half-hour recital by four saxophonists and *Twenty Questions* – Britain itself was at a turning point. There was an end in sight to crippling postwar austerity. New buildings were emerging from bomb sites, the Festival of Britain was in full swing, the Festival Hall having opened three weeks before the first *Crazy People* was broadcast. The airwaves were filled with patriotic programmes like *Our Island Story*, the *Windrush* generation was three years old and rock 'n' roll was just around the corner. Britain may have been on the winning side in the war but the immediate legacy was a curious sense of flux and a sea change in national identity that's still being felt today.

The Goons played a crucial and arguably leading role in this quest for an understanding of Britain's new place in the world. Milligan was a child of the Empire – having grown up in India where his military father Leo spent a large portion of his career the endless grey of Catford's Victorian terraces proved to be a jolt for young Terence when he arrived in south-east London as a 12-year-old. He and his fellow Goons were all in their early twenties when they were discharged from their war service, and when *Crazy People* first went out Britain was beginning to lose its grip over its colonies. India and Palestine had gone and the rest would follow; the Goons' manic subversion of the Edwardian way of life from which Britain was beginning to emerge was a huge factor in the cultural shift that began in the 1950s and which, in many ways, is still churning today. In characters such as Major Bloodnok and the Shere Khan tones of Grytpype-Thynne Milligan had created representatives of that dying imperialism ripe for comic lampooning. The church, the military and the Foreign Office were popular targets for the Goons; they only overstepped the mark when Sellers' outstanding impersonations of Winston Churchill were eventually nixed at the request of the Prime Minister himself.

'I wasn't consciously aware of it but I had had enough of the British Empire,' said Milligan. 'The Goons gave me the chance to knock people my father and I had to call "sir". Chaps like Grytpype-Thynne, with educated voices who were bloody scoundrels.'

The show's audience was broadly a young one – it proved particularly popular among grammar school boys who didn't really fit into a system seemingly designed to manufacture a production line of well-spoken potential army officers and civil servants who could decline Latin nouns at will but not know why. Michael Palin cited the Goons as one of his and the rest of the Pythons' key influences, becoming a fan of the first piece of radio to exploit a generation gap. Part of the thrill of the Goons was in the blank, uncomprehending looks on the faces of the nation's parents (Palin's father even wondered whether there was something wrong with his radio). Rock 'n' roll would come to define and establish the British teenager, but the roots of that earthquake in popular culture lay in *The Goon Show*.

Before the Goons, BBC radio comedy had been timid in its approach. For the first decade or more of the BBC you were likely to hear music hall comics or double acts, but with radio still a fledgling medium most of the nation's comedians were reluctant to divulge their material for a mass national audience when they could tour the same jokes for months or even years and find a fresh audience each time.

In 1938 the BBC introduced *Band Waggon*, the first comedy show specifically designed for radio, although one put together on the hoof: when the Corporation had a whole raft of new dance-band shows ready to go that year, some rudimentary audience research revealed that people were bored with endless broadcasts of the Savoy Orpheans (the eponymous hotel dance orchestra that effectively became house band to the neighbouring BBC) and their interchangeable counterparts. Chirpy Liverpudlian Arthur Askey and urbane Cambridge alumnus Richard Murdoch were drafted in to add laughs, with a premise that they shared a flat at the top of Broadcasting House, but still there remained an adherence to the variety theatres and music halls. When, in 1939, Askey gave up the show because he was too busy with his live work and Murdoch joined the RAF, Tommy Handley was brought in and *It's That Man Again* became a stable of the BBC's wartime fare. When the show began in 1939 it was far from popular and probably would have been canned after a few weeks, but once war was declared its quickfire gags and multitude of catchphrases soon proved enduringly popular and the show ran for ten years until Handley's death in 1949.

Nearly 30 years had passed since the birth of British radio when the Goons, and the visionary Milligan in particular, mortared the entire landscape of radio comedy and changed everything. This was the first comedy series that wasn't merely a variety stage show with a microphone stuck in front of it. It was a show that recognised radio wasn't just a conduit for existing forms of entertainment but a new and magical medium in itself. The Goons became so familiar that it's almost possible to forget how groundbreaking they were: the imagination required to invent that world, the confidence Milligan had to find to not only trust his instincts (not to mention the toll it took on his mental health) but to browbeat the pipe-smoking sports jackets in charge into trusting his instincts too.

It's hard to think of anything on British radio that's been as successful a combination of innovation, edginess and still being genuinely funny to a general audience as The Goons. *Round the Horne* followed, as did the peerless *Hancock's Half Hour*, but they didn't blaze a new trail the same way the Goons did. *I'm Sorry, I'll Read That Again* might be the closest, the first broadcast flourish of the generation that would produce *Monty Python* and *The Goodies*, but even they all acknowledged their debt to Milligan and the Goons. In recent years genuinely innovative comedy on British radio has been a rare thing. Chris Morris's *On the Hour* is an exception yet still maintains little more than cult status even after transferring to television as *The Day Today*. *The Hitchhiker's Guide to the Galaxy* stands out as a true original, and Douglas Adams remained convinced of radio's place as the medium for genuinely innovative comedy right up until his death in 2001. But since then, with only a few exceptions – *The Mary Whitehouse Experience* and *Ed Reardon's Week* spring to mind – it's been the panel perennials *I'm Sorry I Haven't a Clue*, *Just a Minute* and *The News Quiz* that have sailed on while sketch shows and sitcoms have come and gone, too many of them clearly seeing radio comedy as a stepping stone to television rather than a uniquely audio production. It's been a long time since radio comedy did anything controversial enough to have the listening nation dropping plates in the kitchen or upending its gin and tonic into its lap.

'I was never a particular fan of the Goons,' Arthur Mathews told me, 'although it was an iconic show. It just wasn't my kind of thing, really.'

I was quite surprised to hear this as Arthur has produced some of the most inventive and subversive comedy of the last 30 years himself. Probably best known for creating and writing *Father Ted* with Graham Linehan, he also co-writes *Toast of London* with Matt Berry and has worked on a host of outstanding comedy shows, including *Big Train, Black Books*, Harry Enfield, Alexei Sayle and the Ted and Ralph sketches for *The Fast Show*. Born in County Meath, Arthur cut his teeth on Dublin's *Hot Press* magazine where he met Linehan. The pair moved to London in the early 1990s and changed the comedy landscape with *Father Ted* before Arthur returned to Ireland.

'I've always seen Monty Python as kind of The Beatles of British comedy,' Arthur told me via Skype from the kitchen of his house just outside Dublin. 'So I'd say that makes the Goons the Lonnie Donegan of British comedy in that they were very influential on the Pythons. I've always found them a bit too "big", with their silly voices and general over-the-topness. I can obviously see how they must have seemed revolutionary in the fifties, though. Sellers was my favourite. Never liked Milligan much in the Goons, to be honest, but I had definite favourites in the Pythons too: I much preferred Cleese, Palin and Idle to Jones or Chapman. Hancock was good; I've heard a lot of Hancock's radio stuff and it's surprisingly surreal. Him flying jet aircraft; some of it was almost *Python*-like surrealism and very funny.'

Until recently Arthur acted as a kind of comedy consultant to Radio 4, flying over to London from his Dublin home once a month to go through new scripts and advise producers. It was a return to BBC radio for a man whose earliest scripts were broadcast there.

'I was part of a thing called the Programme Development Group that would meet every month to discuss comedy scripts and ideas,' he said. 'Jane Berthoud was head of comedy at the time and she asked me to do it. It was me and a lot of producers, basically. We'd be given these scripts and ideas, and the people who were proposing the scripts would come in and we'd talk about it and see where things went from there. A lot of it was very formulaic, very Radio 4 – a lot of that twee but vulgar stuff they seem to like. You'd occasionally come across unusual stand-ups who people were trying to get a vehicle for in radio. I used to like the offbeat, strange ones. But they weren't very Radio 4 friendly, I suppose.

'One of the first things Graham and I wrote together was a radio show for Lenny Henry – the first thing we did that ever got on air. There was actually a good radio joke in that which went, "I'll jump off this roof – it's only about six feet from the ground." Someone says, "No it's not – it's six hundred feet." "Oh, I never could judge distances." That's a good specific, non-visual radio joke that depends on the audience not seeing what's being described.'

By the mid-nineties, however, Arthur had become firmly wedded to television. He'd met Linehan when they'd both worked at *Hot Press* magazine during the eighties, but the roots of *Father Ted* lay in a character Arthur created during a brief foray into stand-up. When he joined Linehan in London in 1991 the pair turned the character into the focus of a sitcom, pitched it to Channel 4, and the first series of *Father Ted* aired in 1995. There would be two more series before the untimely death of the show's star Dermot Morgan in 1998, and since then television has provided the main focus of Arthur's work.

'People always say the pictures are better on the radio,' he said, 'but I always say, "No, that's not true, the pictures are always better on the television." Obviously for radio you don't have to build a set, so it's infinitely cheaper, but is it more creative? I don't know. I think you can nearly always adapt a TV idea for radio. I've done that kind of thing when a show isn't accepted for TV. I always think that *The Likely Lads* TV show would be more or less the same on the radio because most of it is just the two boys talking to each other. There's very little visual stuff, and I always try when doing something for TV to use the medium for what it's good for, which is visual jokes.'

Despite his television pedigree Arthur is a man long steeped in radio:

'One strong memory I have is listening to *The Glenabbey Show* on RTÉ in the seventies. It was a sponsored show – Glenabbey were a knitwear crowd, I think – but it starred Frank Kelly who went on be Father Jack in *Father Ted* and was very funny. Sponsored shows were a big thing in Ireland then. There was a very famous show, *The Jacob's Show*, sponsored by Jacob's crackers and hosted by Frankie Byrne, an agony aunt who used to play a lot of Frank Sinatra in between solving people's problems. It ran from the early sixties right into the eighties, and later it turned out that for most of it she was having a longstanding affair with another Irish showbiz personality.

'I'd also tune in to the BBC in order to hear the chart countdown on Radio 1, but the reception was terrible because it was coming all the way over the Irish Sea from Britain. I'd listen to *Sports Report* too, again a fairly crackly reception coming across the water. I once taped the commentary of a game between Leeds and Newcastle in 1974, holding a microphone up to the radio, and kept that cassette for quite a few years. It wasn't even a big game, it was an ordinary league match, there was no reason at all to preserve it for posterity. Around the same time a cousin of my father's once wrote a musical that was produced on RTÉ radio and I taped that as well. That was quite a big moment because someone I knew had written something that was on the radio. At one point around then I got strangely obsessed with "Wandering Star" by Lee Marvin. There was a thing on BBC Radio 2 called "power play", where they'd select a particular record every week and play it at the top of every hour. Our radio broke, it was held together with rubber bands, and I found I couldn't retune it to Radio 2 and hear "Wandering Star" on the hour, and I got quite upset about that.

'In the mid-eighties there was a thing on BBC radio called *Son of Cliché* written by Rob Grant and Doug Naylor who went

253

on to write *Red Dwarf*. That was a funny show. They had a great sketch where they did the 1974 FA Cup final as performed by Shakespearean actors. "To thee then, Tosh, and back to thee, and we shall *smite* these Geordie knaves", that kind of thing. Then when I moved to London in 1991 we didn't have a television because we'd just arrived, so I tuned in to Radio 4 and heard Chris Morris doing *On the Hour*. At first I honestly didn't know if it was serious or not, it was so realistic, and to stumble across it like that was terrific. Obviously anything Armando Iannucci does is brilliant. They had the bravura to just go and do news straight, exactly as it was – you have to do something like that fully authentic, going at it even vaguely half-hearted wouldn't cut it.'

Arthur also had a connection to one of the great figures in British music broadcasting.

'I used to listen to John Peel too and tape that now and again,' he recalled. 'I was never one of those Peel obsessives, though. Recently 6 Music re-ran some of his old shows and I found a lot of it horrible to listen to, really. He played some dreary stuff even though he was great, obviously.

'I actually got to know Peel a bit towards the end of his life. I went to his sixtieth birthday party. Also his funeral. He'd got in touch with Graham and I because he liked *Father Ted*, sending us a postcard saying how much he enjoyed the show and telling us to look at the front of the postcard. It was of four scenes in Crouch End in north London. One of them was a picture of a restaurant and sitting there in the restaurant was John Peel. We ended up seeing quite a bit of him because his literary agent was Cat Ledger, who was at Talkback then, the agency and production company we worked through, and we'd bump into him occasionally in the Talkback offices and go for lunch. In fact, the most humbling thing that ever happened to me was when, on *A Good Read* (that Radio 4 show where people picked their favourite book), Peel chose one that I'd written called *Well Remembered Days*. Matthew Parris was the presenter and he was saying, "This book, Mathews, just seems to be having a go at the Irish. Surely it's borderline racist?" And

Peel was defending it, which was a wonderful thing to hear. That must have been around 2001, and he died in 2004.'

I asked Arthur if gets the chance to listen to much radio these days.

'I listen to a lot of radio, more than most people in this day and age, I suspect. I do quite a lot of painting and drawing, and it's great to do that with the radio on. I can't have any background sound when I'm writing, but if I'm doing something that doesn't demand that level of concentration then radio's perfect. The BBC player, in particular, is a fantastic resource. Brilliantly, even the two minutes' silence on Remembrance Sunday is on the player, so if you feel like it you can sit down and listen back to the two minutes' silence.

'Mostly I listen to music shows these days. I love Eddie Piller's soul show on Soho Radio. He managed Jamiroquai and Soul II Soul, runs Acid Jazz records and has all these great stories. That's my main listen, two hours a week. I listen to a lot of 6 Music shows too: Don Letts, Craig Charles and Cerys Matthews. I probably spend quite a few hours a week listening to the radio. It's practically never live, though, mostly it's through the BBC radio player.

'To be honest, I don't listen to much BBC comedy on the radio these days. I liked *Party* with Tim Key in it, a good old fashioned sitcom, and I love The Pin, a British double act who've done a few series for Radio 4. They're great – I'm working with them at the moment on a TV idea. Someone put me on to Colin Hoult a while ago – he's very funny, kind of a sketch show, importantly with an audience. A lot of these radio narrative shows work so much better with an audience; in fact, it's hard to get them to work without one. I'd certainly struggle to think of any that do.'

Ultimately, however, it seems that even for one of the great contemporary comedy writers the alchemy for comic success on the wireless is no great mystery.

'What makes good radio comedy?' he said. 'Funny people, the same as everything else.'

21

The Filthiest Joke in the World: Clapham and Dwyer's Moral Panic

On the evening of 21 January 1935 a statement was issued from Broadcasting House: 'The BBC apologises to listeners for the inclusion in the music hall programme broadcast on Saturday night of highly objectionable remarks violating standards which have been firmly established by the practice of the BBC.' It's not often apologies have to be made for items on comedy programmes, especially radio ones, but so grave was the offence caused by the comedy duo Clapham and Dwyer that Saturday night, and so uproarious the reaction even two days later, that the BBC was moved to issue this official cough and shuffle.

In fact so tumultuous was the moral uproar that the wife of Charlie Clapham, the man whose actions were responsible for the issuing of the statement, and who had effectively gone into hiding, was doorstepped by reporters. She informed them that her husband was 'extremely upset' and 'the last man in the world to say something like that intentionally'.

So what had Clapham said to cause hands to fly to mouths across the land? What had prompted a flurry of calls and letters to the Corporation? What terrible taboo had been broken by the normally amiable and extremely popular giddy half of a pair of comics?

Clapham and Dwyer were the first comedy act to have been made by radio. When they made their broadcast debut in June 1926 they'd only been performing together for a year. They'd met in 1924 in the

256

crypt of the Royal Courts of Justice where Clapham was a clerk to a King's Counsel, while Billy Dwyer was a car salesman caught up in a court case. The two men hit it off and Dwyer, who had done some semi-professional entertaining in his time, realised they were perfect foils for each other. Dwyer assumed the role of a morose straight man, with Clapham his chirpy, rambling sidekick. They were clearly well connected: the first Clapham and Dwyer gig was at a private dinner party given by the Duke and Duchess of Devonshire. Soon after, the society jungle drums saw the pair summoned to Savoy Hill for an audition.

Clapham was reluctant at first – 'We only had one act at the time and didn't want to give it away to the thousands listening in,' he said – but eventually the pair turned up at the BBC, where a nervous Clapham began unintentionally mangling his words to hilarious effect. The pair were booked, and while they may be all but forgotten today they became household names purely thanks to radio.

By the time they appeared on the regular variety bill that night in January 1935 they were veterans of theatre tours and seasoned broadcasters with nine years' experience behind them – marquee names for the Saturday night music hall show, appearing that evening on a star-studded bill with singing sisters Elsie and Doris Waters, Phyllis Robins, A.C. Astor and an act called the Dancing Daughters, whose name suggests their radio appeal might have been limited. At the appointed moment Clapham and Dwyer were soon slipping into their well-established routine of quickfire gags, wordplay and banter. The *Radio Times* billing of their slot as 'A Spot of Bother' would prove to be remarkably prescient.

Clapham's schtick was to wander from the script, playing off word associations with himself, bouncing set-ups off his partner for which he would provide his own punchline. For nine years this had proved extremely popular with audiences and earned the pair a good living.

'What's the difference,' chirruped Clapham at his old sparring partner at the end of a long digression about champagne, 'between a champagne bottle and a baby?'

'I don't know,' Dwyer responded with customary weariness, 'what's the difference between a champagne bottle and a baby?'

'A champagne bottle has the maker's name on its bottom.'

A stony silence filled the airwaves. It was probably only a couple of seconds, but it must have seemed like an entire geological era to Charlie Clapham, standing on one side of the microphone with the blood draining from his face. Bill Dwyer would normally have swept in at this point with some lugubrious critique of his friend's stream of consciousness, but he too was struck dumb. In sitting rooms across the land mouths hung open and couples stared at each other, scarcely able to believe what they'd just heard. Dropped sherry glasses shattered on parquet floors, pipe stems were bitten clean through. A couple of taps of the baton from the conductor and the band struck up a musical sting while the dry-mouthed comics composed themselves and were able to carry on with the rest of their act, albeit in subdued fashion. At the close of the broadcast the ashen-faced comedians went straight to the head of variety Eric Maschwitz and apologised. But they knew it was too late.

Never had such filth been broadcast by the BBC. Not only had Clapham referred to a part of the anatomy that, as far as the BBC was concerned, didn't officially exist, he had also made reference to the fact that babies were created as the result of a physical act that also, as far as the BBC was concerned, didn't officially exist. Retribution was swift: Clapham and Dwyer were banned from the airwaves for six months.

'We have a family of children here,' said Mrs Clapham, 'and you can imagine that jokes of the wrong kind would hardly be acceptable in our home, much less over the microphone.'

'My brand of humour depends on impromptu gagging that I call "padding",' said Clapham when he finally emerged to face the music. 'I thought I could pad the words contained in the script for that particular gag but found I could not. The changed wording turned what was intended as a harmless joke into a really objectionable remark and, what's more, it was not even funny.'

Once they served their penance all seemed forgiven and the duo continued happily until Dwyer was forced to retire through ill-health in 1940 – the early stages of a heart condition that would kill him three years later at the age of 56. Clapham continued the act with a couple of replacement Dwyers for a few years until retiring to the south coast, where he died in 1959 at the age of 65. I wonder what he thought if, in his dotage, he ever tuned in to *The Goon Show*, hearing the anarchy, mayhem and the kind of saucy innuendo that Milligan and co. somehow got away with (a Captain Hugh Jampton, for example) as he looked out to sea from his home on the Marina at St Leonards-on-Sea, barely two miles along the shore from that concrete bunker at the eastern end of Bexhill.

22

'There's nothing between Us and Heaven': the Fleet's Lit Up and So is Tommy Woodrooffe

'This is the Regional Programme. *The Illumination of the Fleet.* Once again we're taking you on board HMS *Nelson* for a description of the scene at Spithead tonight by Lieutenant-Commander Thomas Woodrooffe.'

As an ex-naval man and a versatile broadcaster, Tommy Woodrooffe was the ideal man for the BBC to send to the Coronation Review of the Fleet on 20 May 1937. It was the climax to a week of national high-fiving following the Coronation of King George VI eight days earlier, with the fledgling monarch taking to the royal yacht for a three-day beano culminating in the impressive 'illumination of the fleet' where bulbs were strung along the superstructures of the ships to create spectacular outlines in the dark. On a given signal all the lights would be turned off simultaneously for a few seconds, giving the impression that the ships had literally vanished into the night, before a web of searchlight beams would weave itself into the sky.

Woodrooffe had had a busy week, having been part of the team commentating on the Coronation itself, spending hours in a wooden cabin on Constitution Hill, and then criss-crossing the country reporting from events and conducting vox pops of those enjoying the national hullabaloo. By 20 May he was, like most of the nation, feeling a bit demob happy. In being posted to the battleship HMS *Nelson* to deliver his expert naval analysis of the fleet illumination he was even returning to a ship on which he'd served during his time in the service, meeting old friends and colleagues on every turn he took around the decks. He'd made his first broadcast of the day just before three o'clock, then again two hours later, and was at the microphone at eight, giving an evocative 15-minute description of the scene as the sun set over the ranks of ships, eight lines of vessels stretching prow-to-stern for six miles, including 14 guest vessels from around Europe and the world, and a preview of what was to come later in the evening. A warm day had become a warm night at the end of a long and busy week, and with a couple of hours to kill before his final broadcast of a week of festivities Woodrooffe took off the headphones and capitalised on the opportunity to catch up with a few old shipmates.

Nobody's entirely sure how he spent that part of the evening, but when the London studio handed over to him for the last time after ten o'clock, the Lieutenant-Commander Woodrooffe in front of the microphone sounded a little different from the Lieutenant-Commander Woodrooffe that had delivered the eight o'clock summary. After a brief pause during which the ether crackled and hissed a voice that was at once familiar to listeners yet suddenly unfamiliar began to speak. 'At the present moment,' he said, 'the whole fleet is lit up.'

In black and white, on paper, it's the perfect opening. Simple, succinct and conjuring a visual sketch in the minds of listeners in the space of ten words. There was nothing wrong at all with what Woodrooffe said. What would make the next four minutes go down in radio history, however, was the way he said it.

His delivery was usually authoritative and cheerily brisk, in the manner of a man well used to commentating on a range of events from royal occasions to boxing matches; a man whose job it was to deliver information in as concise a fashion as the occasion demanded. This time, however, he seemed faintly distracted. There was a certain sleepiness in his voice, the words slightly slurred, still delivering in received pronunciation but at 33rpm rather than 45.

'When I say "lit up", he continued, 'I mean lit up by fairy lamps.'

The clip has been played countless times since and it's four minutes of sloshed brilliance, the kind of thing that if you'd heard it in person you'd be catching someone else's eye, raising your eyebrows and waggling an imaginary glass in front of your mouth while enjoying yourself immensely.

'We've forgotten the whole royal review,' he intoned in the manner of a man trying to speak using somebody else's lips before repeating the phrase with more urgency, as if the thought had just properly hit the spot.

'The whole thing is lit up by fairy lamps,' he continued, picking up the pace a little. 'It's fantastic. It isn't the fleet at all. It's just . . .' a pause for a second or two of hiss and static before he blurted out in almost childlike wonder, 'It's fairyland! The whole fleet is in fairyland!'

There was no two ways about it – Tommy Woodrooffe was, at this point, absolutely blootered. In sitting rooms across the land men were taking their pipes out of their mouths and saying to their wives, 'You know, I rather think this chap is as tight as an owl.'

The thing is, if you take away the delivery and look at what Woodrooffe actually says, it's beautiful radio. It's an open and shut case of the magic of radio in action: the conveyance of a wondrous occasion delivered with genuine feeling rather than a dispassionate litany of lazy adjectives. It's repetitive, it wanders, and in terms of being thoroughly stocious Woodrooffe does everything short of belching, and I certainly wouldn't advocate for drunk radio being the best radio, but 'the fleet's lit up' is spontaneous, heartfelt, and in its own way wonderfully eloquent.

'Now, if you'll follow me through, if you don't mind, the next few moments, you'll find the fleet doing odd things,' he continued. 'At the present moment, the *New York*, obviously, is lit up . . .'

This was probably the weakest part of the broadcast. There was not much happening, everyone was just waiting for all the lights to go out, and Woodroffe was struggling slightly and sounding as though he'd been distracted by something. Faintly irritated, even.

'. . . and when I say the fleet is lit up, in lamps, I mean, she's outlined. The whole ship's outlined. In little lamps.'

There were a few moments of silence after this until Woodrooffe returned to explain why he might have sounded a little diverted up to that point.

'I'm sorry,' he said, 'I was telling some people to shut up talking.'

This was the moment that sealed the place of 'The Fleet's Lit Up' in radio history. Telling some people to shut up talking mid-broadcast is probably fair enough if they're proving to be a distraction and there's a chance they might be heard over the microphone, especially on a naval ship where the vernacular might verge on the fruity, but actually informing the audience that he was telling some people to shut up talking left Woodrooffe no way back. Which is a shame, as the rest of the broadcast is absolutely wonderful. 'The Fleet's Lit Up' is destined to remain Tommy Woodrooffe's legacy. That's not necessarily a bad thing because, embarrassed as he certainly was in its aftermath, it is a remarkable piece of radio if you can get past the initial urge to giggle at Mister Boozy. Yet Woodrooffe deserves a little better than being remembered as the old soak who went on the radio with bubbles popping over his head.

Born in South Africa at the start of the Boer War in 1899, the son of a doctor, Woodrooffe arrived in London by steamer from Durban as a 17-year-old in September 1916 in order to take advantage of a special entry scheme for the Royal Navy. He found himself in a city and a nation traumatised by war: it was just weeks since the Battle of the Somme had commenced in disastrous fashion and the streets would have been filled with recruiting posters and stalls and the traumatic sights of the wounded passing through Waterloo and Victoria stations on their return from the front. He spent the four months after he arrived in intensive study at Clifton College to convert his South African schooling into its British equivalent, and a couple of weeks after his eighteenth birthday he joined the navy, starting as a midshipman on HMS *Resolution* at Scapa Flow. 'A South African of intelligence and quick wit,' noted his admission papers, 'he should do well.'

In 1919 he was posted to China where, he told a radio interviewer years later, he saw some extraordinary things. 'One day on the

Yangtse I was sitting on the deck of my gunboat reading *Punch* when I looked up and saw an execution taking place on the shore, some chaps getting their heads cut off by an executioner with a large double-edged sword,' he said. 'It makes an extraordinary *snick* when a fellow's head's cut off, you know.'

He left the navy after nearly 20 years' service having risen to the rank of Lieutenant-Commander and joined the BBC, where he quickly gained a reputation as a versatile broadcaster. One of his earliest outside broadcasts was live from a Romany wedding deep in a Hertfordshire wood, which was shortly after he'd returned from the 1936 Olympic Games in Berlin having commentated on the opening and closing ceremonies, as well as describing some of the live events too, at which he proved to be a natural.

He met Jesse Owens in the Olympic village and asked if he was intending to break any more records. 'He said, "Well, I ain't interested in breaking records, all I likes is running races." And that afternoon he went out and broke two more world records.'

Woodrooffe made such an impression with his dispatches from Berlin he was sent a medal of recognition by Joachim von Ribbentrop a few months later.

'If the BBC finds a chap who's good at commentary they just let him loose,' he said of his versatility. 'They ask what you know about shove ha'penny, you say you know nothing, and they say, "Very well, you've got half an hour to go and find out all you can on shove ha'penny and be ready to broadcast about it."'

Back at Spithead as the moment approached for the lights to go out, plunging the fleet into darkness, Woodrooffe seemed to grow agitated and decided to let the audience in on his thought process. 'In a second or two,' he said, trying to remain calm, 'we're going to fire rockets, erm, we're going to fire all sorts of things. You can't possibly see them, but you'll hear them going off, and you may hear my reaction when I see them go off. Because I'm going to try and tell you what they look like as they go off. But at the moment there's a whole huge fleet here. The thing we saw this afternoon, this colossal fleet . . . lit up . . . by lights . . .'

He drifted off for a second here before a renewed surge of appreciation coursed through him and he bedcame gripped by a sudden childlike wonder.

'The whole fleet is in fairyland!' he cried, almost in disbelief at the vista in front of him. 'It isn't true! It isn't here!'

He reached the peak of his crescendo just as the signal went out for the ships to kill their lights. Woodrooffe went into overdrive.

'It's gone!' he cried with tangible emotion. 'It's gone! There's no fleet! It's, er, it's disappeared! No magician who ever could have waved his wand could have waved it with more acumen than he has now at the present moment. The fleet's gone. It's disappeared.'

It was then that the combination of pressure, excitement and the Pusser's rum sloshing around his bloodstream proved too much. Tommy Woodrooffe let out a bad word.

'I was talking to you,' he said in a more measured voice, trying to rein himself in, 'in the middle of this damn . . .' Realising immediately what he'd said Woodrooffe attempted a sort of combined cough and throat-clear, which might have convinced a few listeners at home that they'd misheard through the crackles of the atmosphere but which would have prompted frantic calls between departments back at Broadcasting House. For now, however, Woodrooffe lumbered gamely onward, pretending he hadn't blasphemed: '. . . in the middle of this fleet, and what's happened is the fleet's gone, disappeared and gone. We had a hundred, two hundred warships around us a second ago, and now they've gone, at a signal by the Morse code, at a signal by the fleet flagship which I'm in now, they've gone, they've disappeared.'

The challenge of describing complete darkness for any length of time is a great one for even the most alert and sober commentator let alone a tired and emotional one, but Woodrooffe then came out with the most beautiful line of the night.

'There's nothing between us and heaven,' he sighed. 'There's nothing at all.'

There was a pause at this point, at which London took the opportunity to intervene and inform listeners that they had to leave the Spithead review and go over a little earlier than expected to the programme of dance music from the Savoy Hotel that was scheduled to follow Woodrooffe's broadcast. A 15-minute broadcast had been truncated to four.

Within an hour the BBC had received around 500 telephone calls from listeners about what they'd just heard. It was an immediate sensation. The following week's live dance band broadcast was punctuated at one point by a male voice crying, 'The fleet's lit up, baby!' near the microphone, while a number of newspapers carried an advert reading, 'The fleet's lit up but see your chickens grow on Kilpatrick's Chicken Rearing Meal, the chickens' breakfast!' If there had been memes in the 1930s Woodrooffe would have been a sensation. Within a fortnight a variety revue opened in London called *The Fleet's Lit Up* and to its credit the BBC broadcast part of it one Saturday night less than a month after Woodrooffe's performance.

Woodrooffe was officially suspended for a week – he was resolute in his insistence that he hadn't been drunk, he'd only had a couple of whisky and sodas; rather he was suffering from nervous exhaustion after an intensely busy period of work – but was off the air for a full three months of what the Corporation called 'sick leave'. The BBC might also have been conscious that during their Coronation coverage a man had walked into one of the commentary positions, picked up the microphone and broadcast for a full two minutes before anyone realised that, far from the expert summariser they'd been expecting, this was just some bloke out of the crowd, making the stern policing of the airwaves a subject of rather thin ice for the Corporation.

When he returned to broadcasting Woodrooffe was busier than ever. On New Year's Eve that year he was sent onto the streets of London to interview revellers and report on the public mood. In May 1938 he commentated on the FA Cup final between Preston North End and Huddersfield Town. Goalless at full-time the game went to extra-time with Woodrooffe the sole commentator. As the game ticked down into its final moments he said, 'If there's a goal now I'll eat my hat.' Preston were immediately awarded a penalty which was converted by their star forward George Mutch. A few days later Woodrooffe appeared live on BBC television where he set about consuming a straw boater made from sugar.

He commentated on boxing and cricket matches, broadcast from the Royal Ballet and described a visit by the King and Queen to Covent Garden. He also broadcast live from among the crowds in Downing Street in 1938 when Neville Chamberlain returned from Munich via Heston Aerodrome, where he'd waved his piece of paper and then from an upper window at number ten had famously proclaimed 'peace for our time'.

'It was raining that day,' Woodrooffe recalled later, 'and I was standing in this enormous crowd with a microphone in one hand and holding an umbrella with the other. It was difficult to hear what Chamberlain was saying up at the window but one did one's best to make sure listeners heard it as straight from the horse's mouth as they could.'

His abiding memory of the occasion was a little more prosaic, however.

'I lost my umbrella that night,' he recalled, 'and said to the BBC, "Well, what about it, I'd like another umbrella." But they said no on the grounds it was lost in the ordinary accidents of life, which I found rather shady.'

When war broke out Woodrooffe left the BBC and went back to the navy. It didn't spell the end of his time behind the microphone, however.

'I also had to do a certain amount of broadcasting,' he said later, 'talking to the Empire about what was happening at sea for five minutes at a time, which sounds quite easy but it wasn't easy because we weren't allowed to say a word about what really was happening. So you filled it up with nice stuff, you know, what might happen if so and so happened.'

After the war he joined the Admiralty, staying for 14 years until retirement.

'I was extremely happy there because nobody knew where you were supposed to be so you could more or less be where you liked,' he reminisced. 'But I was doing something very worthwhile, keeping the navy up to strength.'

Tommy Woodrooffe died in 1978. His obituary in the *Guardian* newspaper tactfully referred to 'The Fleet's Lit Up' by describing him as merely 'lost for words'. As epitaphs go, however, the last line of his infamous Spithead broadcast would have served as well as anything. What better send off could a person have than, 'There's nothing between us and heaven. Nothing at all.'?

23

'Who Cares if the Government's Unhappy?':
Jessie Brandon, Pirate Queen

Britain in the summer of 1984 was not a particularly fun place to be. The miners' strike was in full swing, unemployment was up to 3.2 million and WPC Yvonne Fletcher had been shot dead on a London street outside the Libyan Embassy. The Provisional IRA was planning the Grand Hotel bombing in Brighton that October and bomb scares and evacuations were a regular occurrence in many British cities. There was also the constant lurking threat of nuclear annihilation: Ronald Reagan made his thigh-slapping 'we start bombing in five minutes' joke that summer while in the autumn the BBC screened the nuclear holocaust drama *Threads*, one of the most terrifying pieces of television ever broadcast.

The bus stop and telephone box outside our house were smashed up on a weekly basis, while every house in the vicinity, including ours, was broken into so often that my dad rigged up a complicated device behind the back door involving milk bottles and bits of string that would at least scatter the floor with broken glass when someone forced it open. That spring I saw my favourite comedian Tommy Cooper drop dead live on television while Charlton Athletic, whose cavernous, crumbling stadium in which a team of has-beens and no hopers blundered around the lower reaches of the Second Division was the one place in the world I felt I might properly belong, were so broke they came within five minutes of being liquidated in the high court.

One evening that summer I was moving slowly through the ether, passing along a procession of French torch songs, German news bulletins and a Dutch orchestra, towards the farthest point on the left of the medium waveband. There wasn't usually much that far along the dial, but just as I passed 600kHz and was about to switch off there was a sudden burst of a familiar song, Nik Kershaw's 'I Won't Let the Sun Go Down on Me', a particular favourite of mine, so I left it on. As the track faded out there came a series of strange sounds, high-pitched, descending electronic stings, the sort of noise a ray gun would make in a science fiction B-movie. Beneath them Nik Kershaw segued seamlessly into 'White Lines (Don't Do It)' by Grandmaster Flash and Melle Mel, another song I liked, so I took my finger off the wheel and sat back and listened. I didn't know what this station was but I knew I liked the records they were playing. I was curious to hear the language in which the DJ would back announce the song too, thinking French was a good bet because it was a strong signal, so the transmitter couldn't be too far away even allowing for the natural boost radio waves receive at night. The last thing I expected to hear was the smoky, languid voice of an American woman.

'That's Grandmaster Flash and Melle Mel with "White Lines (Don't Do It)", she said. 'This is all hits, all the time, Laser 558 coming live from the North Sea on the MV *Communicator* where the music is never more than a minute away.'

After Kraftwerk's 'Tour de France' the same drily unhurried voice identified the song and added, 'Twenty-six past the hour, this is Jessie Brandon on all-Europe radio, Laser 558, rocking and rolling on the North Sea with Frankie Goes to Hollywood.'

Famously, 'Relax' had been banned by the BBC due to what Radio 1's breakfast DJ Mike Read had decided was the overtly suggestive nature of its lyrics, so hearing it on this mysterious new station provided a *frisson* of illicit excitement. I turned the volume down a bit in case my mum happened to pass the door and cop the salacious lyrics that I didn't understand, but this was radio like I'd never heard before. It sounded fresh and exciting, a station and a DJ that seemed entirely untroubled by the hell-bound handcart of Britain in the summer of

'84. This was a place where the music came first, with a minimum of vacuous prattle. There was no pompous unilateral banning of records, no cheesy links full of crappy catchphrases, no transparent plugging puffery, no adverts and certainly no flipping 'Our' flipping 'Tune'. 'This,' I thought, 'is how music radio should be done.'

Laser 558 was a new pirate station broadcasting in the great tradition of the sixties from a ship in the North Sea, banging out a daily dose of impossible glamour to light up some austere times. The DJs, all of them American, sounded unutterably cool, even though the never-more-than-a-minute-away music policy meant their links were kept to a minimum. Their very names – Jessie Brandon, Tommy Rivers, David Lee Stone, Rick Harris, Charlie Wolf – sounded a world apart from the prattling Phils, Mikes and Steves who normally populated the BBC and commercial radio airwaves. In addition there were jingles billing the station as 'all-Europe radio' and I was thrilled by the idea there were kids like me in Sweden, Austria, maybe even as far as East Germany or Russia on the other side of the Iron Curtain, stumbling across this same magical source of transatlantic glamour beaming from the North Sea. With its pirate notoriety even earning column inches in my dad's newspaper, listening to Laser felt like a minor act of rebellion to a 13-year-old kid.

Laser would only last for about 18 months, but the energy and glamour it provided from its gravity-defying transmission mast on the deck of the *Communicator* was an absolute lifeline to me. Laser was a portal to a world in colour rather than the monotone suburban grey of my every day. Also I'd found Laser myself, not seen an advert or heard about it at school. It was something cool, something new, something I could actually call my own. I'd listen whenever I could and even today when I think of some of the hits of the time – 'Two Tribes', 'When Doves Cry', 'Eyes Without a Face', 'Dr Beat' – they play in my head with the echoing boom and static underlay of the lower end of the medium waveband where Jessie Brandon and Charlie Wolf stoked in me a nascent and unfamiliar sense of self-worth.

There had been plenty of other pirate stations over the previous couple of decades and many that lasted a lot longer than Laser. On the FM waveband at that time there was the likes of JFM and London Music

Radio – indeed for significant chunks of the eighties and nineties you'd struggle to find an inner-city tower block that didn't have an aerial on the roof and kids in a flat banging out soul and reggae records to the locality. Laser was a huge international operation, though, bringing something different and original to British ears, and for a few glorious months it put a little bit of much-needed transatlantic pizzazz into the country while giving British radio a much-needed kick up the arse.

The roots of Laser date back to the pirate radio boom of the 1960s. The explosion of pop music and the birth of the teenager were woefully neglected by the BBC on all three of its national networks, the Light Programme, the Home Service and the Third Programme. Hence in the mid-sixties various entrepreneurs, geniuses, chancers, romantics and charlatans began setting up alternatives, eyeing gaps in the market and the wavebands that only Radio Luxembourg was attempting to fill. The North Sea was the perfect place for a radio revolution: most of it was outside British territorial waters and hence outside the strict jurisdiction of the Postmaster General, providing the entirely legal possibility of broadcasting unregulated over a wide area without the fear of PC Plod pulling up outside in a panda car. There were even some abandoned old forts sticking out of the water where radio stations could base themselves, as well as the ships that were fitted out with transmission equipment before sailing for the horizon to bring pop to the masses. Radio Caroline was the first of these British offshore pirates, launching in 1964, and within three years the number of pirate stations serving the British airwaves was in double figures and pulling in an estimated 15 million listeners. The pirate threat was so tumultuous that it led, in September 1967, to the biggest shake-up in the history of British radio. The BBC re-launched its national services with the Light Programme becoming Radio 2, the Third Programme Radio 3 and the Home Service Radio 4, with Radio 1 serving the pop market with a roster of DJs culled mainly from the pirate stations. Indeed, the first voice ever heard on Radio 1 was a former Radio Caroline and (pirate) Radio London DJ called Tony Blackburn.

In the summer of 1967 the Marine Broadcasting Offences Act was passed, designed to close as many loopholes in the Wireless Telegraphy

Act of 1949 as possible in an attempt to cripple the offshore pirates. It became an offence for British subjects to work for the stations, and also for any supply tenders from British ports to service the ships and forts. With their only source of income – advertising – also cut off by the act, the pirate stations soon packed it in and when Caroline saw its vessel repossessed for unpaid bills it looked as if the age of offshore pirate radio was over for good.

Caroline wasn't done yet, however, returning in the early seventies on board the MV *Mi Amigo* and operating just off the Dutch coast. The early days of the re-launch were blighted by funding issues, not to mention the ending of the BBC's radio monopoly with the dawn of independent commercial radio in 1973, but the station's album-oriented philosophy meant it wasn't pitching itself as a direct rival to the likes of Radio 1 and Capital, while enabling the station's charismatic founder, Irish businessman Ronan O'Rahilly, to expound his mystical hippy brand of Eastern philosophy he called 'loving awareness'.

When the Dutch government also introduced legislation aimed at cracking down on the pirates off its shores Caroline's logistical problems increased, the *Mi Amigo* being forced further out into the North Sea making it harder and more expensive to keep running. One night in the spring of 1980 the ship lost its anchor in a storm, ran aground on a sandbank off Sheerness, began taking on water and left the DJs and crew needing to be rescued by the RNLI before the vessel sank. Its aerial stayed visible above the waves for years afterwards, looking for all the world like a tombstone on the watery grave of offshore radio.

But still Radio Caroline refused to die. In 1983 the station re-launched again in a blaze of publicity on a new ship, the MV *Ross Revenge*, circumventing the 1967 Act by supplying the ship from faraway Spain, which permitted offshore broadcasting, and promising a thrilling modern output and a bright future. But a confused attitude to programming made the station a shadow of its former self and left the seas open to one new radio venture in particular.

As Caroline attempted to reinvent itself the seeds of Laser were being sown. Although overseen by a New York airtime brokerage company Laser's ownership was shrouded in a financial fog worthy of the North

Sea in autumn. The new venture's mystery backers were believed to be based in Ireland, leading to crackpot rumours the station would be a front for the IRA. A German-built former cattle transporter ship was purchased, taken to the Florida Everglades, fitted out for broadcast, renamed the *Communicator* and, in the spring of 1984, set off across the Atlantic with a crew of fresh-faced young American DJs and promises of hefty sponsorship from major international heavyweights. After stops in the Azores and at New Ross on the south-east coast of Ireland, the *Communicator* sailed up the channel, into the North Sea, made sure it was safely outside UK territorial waters, dropped anchor and began test broadcasts as Laser 558. The *Communicator* was staffed entirely by Americans and supplied from the Netherlands, safely circumventing the 1967 Act, and after a few minor early crises including the arrest of the men involved in the clandestine construction of the transmitter aerial in the wilds of Kent, Laser 558 began broadcasting on 24 May 1984. Within a month the new station was claiming a startling listenership of eight million, rising to nine million a few weeks later. These figures represented an almost unbelievable achievement: with Laser not allowed to advertise or promote itself in Britain its audience was garnered almost entirely by word of mouth.

One advantage the station had, at the core of its ability to boast that music was never more than a minute away, was that being located in international waters it wasn't bound by the ancient 'needle time' directive that hamstrung the legal British BBC and independent stations by restricting the amount of recorded music they could play to only a few hours in total every day. This left Radio 1 able to play at most about eight records an hour (a startling figure, and the reason why John Peel began his legendary sessions, recorded as live to circumvent the archaic directive). At three minutes a track that meant more talk than music unless artists played live (a state of affairs that lasted until the abolition of needle time in 1988). Hence Laser's American-style, hit-orientated, minimal prattle conveyor belt of 45rpms was fresh, invigorating and caused an absolute sensation. It also meant they were struggling for advertising. Philip Morris tobacco had agreed a sponsorship deal early in the station's preparations but pressure not to advertise on Laser from European governments meant it never materialised, and it would be months before the station broadcast any significant advertising at all.

While the audience basked in the reflected glamour of this new exotic transatlanticism coming from the choppy brine of the North Sea, life on board the ageing freighter with nothing between the ship and the freezing winds blowing down from Scandinavia probably wasn't a whole big bumper bunch of fun. The broadcasters worked one month on and one month off, with their quarters a basic cabin measuring nine feet by six. I remember one day hearing Jessie Brandon referring to a force-ten gale before announcing, 'This is Jessie Brandon checking out. We've got a few weather problems out here, as you can probably hear, but, hey, stick around your radio, we'll be back as soon as we can,' and the air going dead. For two nights I scanned the airwaves anxiously but found nothing, until eventually the *Communicator*'s storm-damaged transmitter was repaired, the seasick DJs could finally leave their bunks and the station returned to the airwaves.

By August 1985 Laser's popularity had become so great that the British Department of Trade and Industry chartered a ship called the *Dioptric Surveyor* to monitor both the *Communicator* and Radio Caroline's *Ross Revenge* and attempt to block any vessels bringing supplies to the pirate stations. DJ Charlie Wolf christened it the 'Eurosiege'. In September a song was released in support of Laser called 'I Spy for the DTI' under the band name The Moronic Surveyors, but by October the blockade was causing more and more output blackouts amid reports of fuel shortages and a crew growing less happy by the day. Eventually, at lunchtime on 5 November, after a night of heavy storms and halfway through 'Oh Sheila' by Ready for the World, Laser 558 went off air for the last time. The *Communicator* was escorted by the DTI to Harwich and impounded on behalf of the station's creditors. The ship passed subsequently through various radio hands, mostly in the Netherlands, until it ended up in Orkney where in the mid-2000s the old warhorse was scrapped.

I wondered what had become of Jessie Brandon, the coolest voice of my youth. I vaguely remembered something about her moving to Capital after her spell at Laser but couldn't be sure. An internet search didn't reveal much, but it did give me enough information to fire off a speculative email to a radio station in the USA where a woman of the same name had a weekly show. I asked if she was *the* Jessie Brandon and might be happy for me to quiz her on those far-off days of radio piracy. A few days later a reply came back. 'Sure,' it said. 'That'd be fine. Best, Jessie B.'

There's a film effect called the dolly zoom, a piece of visual wizardry in which the camera lens zooms out while the camera itself is pushed forward, completely distorting the perception of the viewer by apparently zooming in and panning out at the same time. Hitchcock used it in *Vertigo*, Spielberg in *Jaws*. When I read that short reply from Jessie Brandon time itself seemed to simultaneously shorten and lengthen. I had my own personal dolly zoom, feeling at once firmly in the twenty-first century yet also 13 again at the same time. I rattled off a list of questions and a couple of weeks passed until one morning I woke up to an email in my inbox. Its subject line was 'Here are my thoughts, let me know if you need anything else' and the email was blank save for an mp3 sound file titled 'Charlie'. I opened it up and the mental dolly zoom twanged my subconscious again.

'Gooood morning, Charlie,' breathed the same laconic voice I'd picked up by chance that night in 1984. 'Let's see if we can do this audio because I talk sooo much better than I type.'

Jessie's voice was crystal clear, just her and a microphone, no jingles, no music, no time checks, no medium-wave atmospherics, no big aerial on a wave-tossed ship, no interruptions. The woman whose voice effectively opened up the world of radio to me had sat down in a room and told the story of a magical radio life through the only medium she knew how: the microphone.

'My first radio memories are the same as anybody's around the world,' she said.

'Underneath the covers late at night – in my case in Washington DC. When I was a kid there were clear channel stations, AM stations that were on wavebands protected from interference by other stations that, after dark, would just broadcast across our side of the Mississippi. There was WOWO in Fort Wayne, Indiana, and KDKA out of Pittsburgh – in the US the "W" stations are east of the Mississippi and the "K" stations west – and they always had the new Beatles records first. Why they put them out at midnight when they knew most of the listeners were kids I can't tell you, but there I was, under the covers, getting the A-side *and* the B-side of the latest Beatles single. There were also the chart shows, which you'd write down by hand, calling your girlfriend and asking, "Did you get number 47? I had to go to the bathroom." Then, as I got older, progressive radio started taking over after hours on the new FM band.'

In the US progressive stations emerged around the turn of the seventies, playing progressive rock and giving a left-leaning slant to the news at the time of the Vietnam War and the civil rights movement. 'FM had the quality to do justice to progressive music,' said Jessie.

'So, yes, I started out top 40 and then grew up into hippie stuff. In my teens I went on a camping trip with my sister and our boyfriends. Her boyfriend said to me in the middle of the night, when were up talking in the dark, "Jessie, you should be on the radio." I thought, "Wow, OK, that sounds great to me." I was studying journalism at the University of Maine at the time, but I switched my major to TV, radio and film, moved back home to the DC area and got my degree at the University of Maryland where I volunteered at the student radio station and after that spent a couple of years working at a leading rock station in Seattle. I'd had my own massive #MeToo moment there, so came back to DC from the west coast, worked part-time for a couple of years, and then saw this ad. A new station in Europe wanting DJs. Hmm, OK, see the world, have an adventure, that appealed.'

Having landed a job with Laser Jessie travelled down to Florida to prepare for the transatlantic crossing.

'When I got to Florida, there were a whole lot of management problems,' she recalled. 'Lots of struggles for power. The programme director who hired me was fired, so they asked everyone to submit a proposal as to why they should be the boss. I was the only one who didn't apply, so they gave the job to me.'

The stop on the southern coast of Ireland was to allow the *Communicator* to make its last preparations before parking itself in the middle of the North Sea.

'Ireland was a riot,' Jessie said. 'Our cover story was that we were something to do with oil rigs and the DJs were supposed to be the scientists. Yeah, right!' She chuckled again. 'They set us up in a hotel in New Ross, the actual seamen stayed on the ship and somebody came from the local paper to interview us and we ended up on the front cover – as scientists. We left a week later and then the truth came out and it was a bit embarrassing for those poor folks. But I had a great time in New Ross, it was right before Christmas, wonderful people.'

Technical issues – the original plan to attach the station's transmitter to a helium balloon above the ship proved unworkable as soon as the first gusts of North Sea wind hit the vessel – meant there were five months between the *Communicator* dropping anchor in the North Sea and Laser taking to the airwaves. I'd asked what Jessie remembered of 24 May 1984, Laser's official first day of broadcasting.

'Not a damn thing – we had a lot of champagne. To me it was a case of, "Oh, *finally*, we're normal." Because to me normal is broadcasting, and I was tired of the waiting. Now it begins. We had a full support staff when we first got on the boat but that was soon reduced to a skeleton crew. We had a guy who was there to cook for us but we took it in turns. Everything was very informal, just me and a whole bunch of guys, on a boat, each of us trying to avoid the people we weren't getting on with that week. It was not fun a lot of the time and you looked forward to getting off the boat, hoping there would be money waiting for you when you did. There usually wasn't, but, hey, at least you were off. We weren't aware how popular we were at first – you weren't aware of anything much until you got off the *Communicator*. But that part was fun, you'd go off and stay in Bayswater on your shore leave and Laser's blasting out of every luggage shop and hair salon on the street – it was great.

'I think Laser was successful because of its informality. My theory is that it's tied in with the difference between the histories of radio in Britain versus America. In America the civilians took over the airwaves from the start then the government eventually stepped in to straighten things out, whereas in England the government had all the hardware and doled out the frequencies so you worked with the government, meaning there wasn't that same informality to British broadcasting. Also the needle time thing, oh my lord, that was the bottom line, wasn't it? Y'all had to stop, even the most progressive DJ had to slow right down and say, "That was . . . and then he'd take a breath . . . and this is . . . The Beatles" and then there'd be another pause. I only learned about needle time when I went to Capital and had to keep pausing, because dead air didn't count towards your needle time. At Laser we hit the airwaves and did it the American top 40 way: "And now here's the Beatles, BAM!" You don't have to wait so that you don't end up owing performance rights money, especially when you're a pirate too.'

There was that throaty chuckle again. 'So that contributed to the success: we were informal, we slouched around, we came from a culture of being informal. I always felt that British jocks said, slowly, "I want you to listen to this . . . and here it comes . . ." whereas we said, "HEY! Get a load of this!" And then we'd hit it, you know? It was a totally different style.' Were you conscious that the British government was unhappy, I'd asked? She chuckled again as she read out the question to herself.

'Who cares if the government's unhappy? I've kept my passport from that time: it shows me *leaving* the UK, but officially I never entered it because I'd entered the country illegally by boat. It was all very subversive: we were there to *make* the government unhappy, to make waves, so it amused us very much. We did have some fear of getting caught, though. I remember going on my first vacation, leaving England but with nothing in my passport to say I'd come in. There I was, waiting in line to get my passport stamped to go over to France. The guy started asking some questions and one of the other Laser folk was in the line behind me and I could feel her just quietly fading away and moving to another line. I got through, though, they didn't mind too much. I'd been told to say I'd come in on a fishing boat, or something.'

Once Laser was up and running more women DJs arrived, but in the early days Jessie was the only woman on the air and, indeed, on the *Communicator*.

'Women DJs were rare then and rare afterwards for quite a while in Britain,' she sighed. 'Ireland too: I did a spell at Radio Nova in Ireland for a while where the only other woman in the schedule did a cooking show. In America in the seventies I was usually the first woman and usually the only woman at the station, and that went on for a good decade or so. There was a theory that people didn't like to hear women's voices, so you couldn't put them next to each other in the schedule, you had to have a buffer of several male DJs between them. That was a real drag and, of course, it was all bullshit.'

After a few months Jessie disappeared from the Laser airwaves, and I'd never found out what happened.

'I think Capital bought me from Laser, I never actually got the whole story. I was on shore leave and received a communication out of the blue from New York that Capital wanted me to work for them – would I be interested? It was arranged that they'd come and talk to me, and they had a car waiting and the whole bit, so I think I got bought, basically. I had some problems with the work permit and couldn't broadcast, but it was written into the paperwork that they had to pay me while they tried to get me my work permit, so I got to go home to America for a while and all that time they were still paying me. In the end they set me up in a studio in New York for three months, and once a week they'd fly my producer Declan

Meehan over and we'd do the American chart show live on Saturdays, then we'd have lunch, tape the Sunday show and then Declan would get back on the plane and take the Sunday show back to London with him. They managed to get me a work permit that covered those two shows, so I moved to London and lived there for a year, but then it was turned down. The Home Office decided I was not in a specialist category and any British DJ could present the American chart show, so Jessie was just a cog in a wheel. I worked at a station in Dublin for about a year, went home to the US and a rock station in Raleigh, North Carolina, then somebody called saying Luxembourg wanted me, so I went back there for a year. It was fun and I saw a lot but Luxembourg was kind of a boring place to live. At the end of that contract I called the Home Office to see if I could go over to Britain and do anything at all. The woman I wound up talking to was actually the one who'd made the original decision and she just said, "Nope, sorry, that was me and I'm not going to change my mind." I went back to the States and that's where I've been ever since.

'I did Smooth Jazz for a while in DC, then I did some news, but my mother fell ill with dementia. I took care of her and did news shifts in DC, and that's what I was doing on 11 September. Then I decided I wanted to work with plants, so went back to university and took a degree in horticulture. When my mother passed away I decided to go back to Maine, where I'd started college, and got a job at a greenhouse. I had an old friend at a radio station there so started work for the Portland Radio Group [in Maine]. They've got seven stations and right now I do shifts on three of them, generally weekends and fill-ins. It's a lot of fun. I get a lot of freedom to be a smartass a couple of times a week. Away from that I get to work out in fields with things that behave properly.'

That chuckle again.

'As for the future, well the internet is the future. It confuses me all the discussions people have in Britain and Ireland about AM versus DAB: the key thing is the reception and the receptor. The audience now is listening in the way they want to listen, on the internet. You can have some fancy aerials and equipment and get the old stuff together and broadcast that way but eventually there's going to be nothing there. No one's going to have any inclination to listen to you if it's going to be in any way difficult to do it, so yes, the future is the internet, period. Every station has a listen now button online, the Tune In radio app means you can get it on your mobile phone; Alexa, the Amazon Echo, you can direct it to play on your furniture. The internet's the future. I don't know why anyone fools around with other stuff any more.'

I'd ended my email by asking Jessie what she thought made good radio. As far as I was concerned she'd produced some of the best radio I'd ever heard, bringing a new freshness to the medium, her dry wit and faintly world-weary delivery unlike just about anything available to British listeners at the time and coating a sheen of American glamour over an eighties Britain where life was, in general, grim. Jessie and her Laser colleagues were like a light in the dark. People tend to attach a sense of ironic nostalgia to being a kid in the eighties – the clothes, the hairstyles, the hair products – but it was a dull era in which to grow up. Laser taught me that life could be fun, and it taught me it was OK to be an outsider. It was all right to think Mike Read, DLT and the rest of the mainstream were terrible, because other people felt that way too. Radio in my city didn't have to be just the BBC, Capital and LBC; Laser 558 made me fall in love with radio and started a lifelong relationship with the medium that has enriched me in countless ways. Accidentally stumbling across Jessie on the airwaves that night in the summer of 1984 literally changed my life.

'"What makes good radio?"' she read. There was a pause as she exhaled. 'Well, as far as what the jock can control, you've got to give in to the magic. You are right there, you are right in the person's head, the person you're talking to. They're a person, they're your friend, not a "listener", and that's how it's got to be. You have to give in to that. You can't pretend.'

24

'Pack it in, Beatrice': Revisiting the Cello and the Nightingale

It was a dark and blustery morning. The wind rattled the eaves and tugged at the window frames and the lights began to flicker. Sandee put down her coffee cup and raised her eyes towards the ceiling.

'Oh, pack it in, Beatrice,' she snapped. Immediately the lights stopped flickering.

Of all the radio locations I was visiting this one was possibly the most intimate and one of the most historically resonant. Among the BBC's earliest broadcasts – the constant round of talks, dance bands, classical concerts and readings – one stood out from the rest. It took place late in the evening of 19 May 1924, and it took place yards from where I was sitting in a kitchen in a converted Tudor barn in rural Surrey.

Beatrice Harrison was in her late twenties when she moved to Foyle Riding, near Oxted in Surrey, in the early months of 1923. The old barn had been converted into accommodation for guests and would serve as Harrison's music room; it dates back to the sixteenth century, and its galleried central room was – and remains – perfect for recitals and performances given by the light of the fire in the vast fireplace. The barn's current owner and resident Sandee Brown, an events organiser of great dynamism and charm, rents out part of the barn to guests, and during the summer stages music recitals in the music room and out on the lawn behind the building, just as Beatrice had done.

When she arrived at Foyle Riding Beatrice Harrison was already a classical music superstar and a particular muse of composer Frederick Delius. Harrison premiered his cello sonata in 1918, while the composer dedicated his 1915 double concerto to Beatrice and her violinist sister May. In 1921 Beatrice was the soloist for the premiere of Delius's cello concerto, and so taken was he with the musical Harrison sisters that he's buried up the road in their local churchyard, a few yards from Beatrice's grave, at his own request. Elgar was a fan too: whenever he personally conducted his cello concerto he insisted on Beatrice Harrison as his soloist.

'I believe Elgar and Delius spent a good few nights here,' said Sandee, raising an eyebrow as she lifted her coffee cup to her lips.

Almost as soon as she'd settled in at Foyle Riding Harrison began taking her cello outside to practise and, late one evening in May 1923, she was sitting at the edge of the woods sawing away at Rimsky Korsakov's 'Chant Hindou' when she heard a nightingale echoing her cello with its song. Enchanted, for a few minutes Harrison played short phrases, for each of which the nightingale trilled its accompaniment. The same thing happened over each of the next few nights until the nightingale had established its territorial claims and fell silent for the rest of the summer.

The following spring the nightingale returned and one night after Harrison had performed Elgar's Cello Concerto live on the BBC, she mentioned its presence to the station's announcer, Rex Palmer. Encouraged by his excitement, she called the BBC and managed to get through to John Reith himself, suggesting he send some boffins and equipment down to Oxted and broadcast a duet between her cello and the nightingale live to the nation. Reith was understandably reluctant at first but he knew Harrison to be one of the country's most famous musicians and, after enduring a few minutes of persuasive hectoring from the other end of the line, commissioned a BBC team to head to Foyle Riding. They arrived equipped with some batteries, an amplifier, a reel of cable, a brand new Marconi–Sykes magnetophone – a revolutionary microphone commissioned by the BBC that provided better audio quality than any previous broadcast microphone – and an air of bafflement and anxiety regarding the task they'd been dispatched to perform. Given the unprecedented nature of the broadcast and the opportunity it presented to test the magnetophone in the wild, Peter Eckersley himself was one of the trio of BBC men who came down to Oxted to set up.

'Here's a photo, taken out there in the porch,' said Sandee, handing me a frame in which the BBC men were arranged on the benches under the gable roof outside the front door, two engineers in headphones looking anxious as they adjusted the equipment, Eckersley sitting opposite them with a telephone receiver pressed to his ear – the broadcast would be sent via the Harrisons' telephone line to the BBC headquarters at Savoy Hill and then transmitted to the nation.

Having tested the equipment the evening before the broadcast – and found the new microphone so sensitive it even picked up the snuffling of rabbits and scuttle of rodents – out went Harrison into the garden at 9 p.m. to warm up. The nightingale's previous appearances had been fairly late in the evening, so the plan was to set everything up ready to go, have Harrison start playing and if and when the nightingale joined in the BBC would cut into the Savoy Orpheans concert being broadcast at the time and present the performance from Oxted live.

By 10.45, with barely 20 minutes to go until closedown, there had been no sign of the nightingale and Eckersley and his colleagues were beginning to exchange glances, but then, as Harrison tried the 'Chant Hindou' again, the bird finally burst into song. Word went back to London, the Savoy Orpheans were faded out, the announcer handed over to Oxted and the listening audience of around a million were transported to a Surrey garden where Harrison played the 'Londonderry Air' while the bird chirruped its accompaniment. For 15 minutes the nation was spellbound by the extraordinary sonic combination of cello and nightingale. The bird's song was captured perfectly alongside the mournful tone of Harrison's cello in the woods, playing barely 30 yards from where we sat at Sandee's kitchen table.

It might not be too much of a stretch to claim that broadcast as one of the key cultural moments of the twentieth century. There was one of the country's finest musicians – as advanced a performing artist as had ever lived, at the peak of her powers, playing some of the greatest melodies ever created by human beings (from elite composers to folk tunes with origins lost in time), and playing them entirely in concert with one of the most beautiful sounds in the natural world. Not only that, the apogee of human technical development was on the premises to disseminate the sound of this extraordinary encounter to people in a radius that reached hundreds of miles. The pinnacles of three key aspects of the world in which we live – creativity, nature, science – intertwined to create one of the great radio and cultural moments of the century.

The reaction was like nothing the BBC had ever known. There were 50,000 letters in response, sent from locations as far afield as Scandinavia and Spain, places where the song of the nightingale had never been heard. The performance was released as a gramophone record and repeated the following year to similar acclaim until it became an annual event, a cornerstone of the broadcasting calendar. Every spring until 1936 Harrison would take her cello out into the woods, start to play, and before long the nightingale would respond.

In 1942, in an effort to boost morale during the Second World War, the BBC returned to Foyle Riding to record the nightingale without a cello accompaniment. Sure enough, the bird began to sing. But suddenly there was a low drone, faint at first but growing louder and more menacing. The engineers realised it was the sound of aircraft – lots of them. What they could hear, and what the nightingale obliviously sang over, were 197 RAF bombers passing overhead as darkness fell, on their way to bomb the German town of Mannheim.

It's an eerie sound even today – there are nightingale recordings of Beatrice Harrison and the bomber squadron easily locatable online – and the engineers were still in the garden when the squadrons returned, recording the same sound only 11 lost aircraft lighter.

Before I left Foyle Riding I stood at Sandee's back door and looked out towards the trees where Beatrice Harrison had conjured the song of the nightingale from the branches to create one of most spellbinding moments in the history of radio. All I could hear was the hiss of the heavy rain on the grass, then the lights flickered again in the storm.

25

'The World is a Small Place When You Have a Radio': Sunday Morning with Cerys Matthews

Cerys Matthews is coming to the end of her three-hour Sunday morning radio show on BBC 6 Music. She's playing a live version of 'Myfanwy' by the Welsh entertainer Ryan Davies, recorded in the mid-seventies and introduced by Davies himself as 'the greatest love song that has ever been written in any language – you name it, Russian, French, German, Spanish, Dutch, anything you like'. It's a nineteenth-century song with roots in a fourteenth-century poem, and a repertoire staple of Welsh male voice choirs. The clock on the studio wall is ticking towards one o'clock and Cerys's producer Angela is on her feet collecting together the pieces of paper on which her colleague Jonny has been delivering texts, tweets and emails to the presenter during the show. In a seat next to me Jonny is going through the final messages before disappearing into the production offices and returning with two bulging carrier bags filled with the CDs and books sent in that week for the presenter's attention that she'll take home with her. There's an end-of-shift feeling: the school bell's about to ring, the factory hooter about to sound, but Cerys is still in her seat with the song pumping out of the studio speakers, her eyes closed, singing along, arms waving in time to the song's stirring crescendos.

Over the previous three hours she's played a Rod Stewart album track from 1970, Irish folk legend Christy Moore, septuagenarian Turkish singer-songwriter Selda Bağcan, British-Asian jazz clarinettist Arun Ghosh, Ethiopian accordionist and Washington DC taxi driver Hailu Mergia, Stormzy, Dr Dre, Nina Simone and Curtis Mayfield. She's featured some little-heard George Harrison songs to mark what would have been the former Beatle's 75th birthday. In the live room next to the studio there's been a session from Liverpool band The Fernweh. There's also been poetry, an author in to talk about the imminent arrival of spring and its effect on the tides and the stars of the night sky, even a vintage recording of the Irish singer Joe Heaney describing the best way to fish in Connemara.

Eclectic doesn't even begin to cover it, but if there's one thing that unites this whistle-stop global odyssey of music and speech it's heart. All humanity is here, humanity at its best, expressing itself in multifarious musical styles drawn from a range of eras. It's a whirlwind of voices, instruments, genres and stories, with Cerys Matthews as its centre.

For a station whose last rites were being read within the last decade BBC Radio 6 Music is looking perkier than ever. In 2010 a Damocletian sword seemed to be hanging over the network when the BBC itself recommended its abolition on the grounds of cost-saving and because the national broadcaster, according to its official announcement, 'recognised the lead role that commercial radio plays in serving popular music to 30- to 50-year-old audiences'. A huge and vocal campaign protested the proposal and led the BBC Trust to reject the idea. The surge of goodwill that helped to ensure the future of the digital platform was followed by an increase in its listenership that hasn't stopped since and meant that within two years of its threatened closure 6 Music's listening figures overtook those of Radio 3. Now, celebrating overtaking Radio 3 in the listenership stakes may seem a little like celebrating breasting the tape in the 100 metres hurdles ahead of Donald Trump in a taffeta ballgown – something I'd pay to see, incidentally – but 6 Music is a digital-only station while 3 can be accessed via both digital and analogue. 6 Music is one of the radio success stories of the millennium and Cerys Matthews' Sunday morning show is a huge part of that.

'The internet's changed everything,' she tells me as we park on a violently pink sofa in a corner of the studio control room.

'As a child I was always listening to older stuff because in the eighties I didn't like the *pew pew* electronic sounds of music then, the shoulder pads, the songs about money and power and material worth. I wanted to know more about the *stories* of the songs, the playing, the composition, the people behind them. I had a huge collection of songbooks from different cultures, including an Irish songbook I was given when I was very young, so I learned about Irish history and particularly the tragedy of the Great Famine from these incredible songs that gave me a sense of the power of music. Unfortunately commercial radio quite often doesn't touch the kind of taboo topics that you get in folk songs, in particular, songs that aren't as commercial. There are exceptions – the Temptations, Marvin Gaye, Sam Cooke – but on most mainstream music radio you're just hearing songs about love, and their programming is a way of profiteering. Now, with the internet and digital radio, it's a different story, you can find what you want to hear more easily. One of the great things about radio is that when you do find that particular something it becomes great company. I go online and listen to jazz straight from New Orleans on WWOZ for example, or Fip Radio in France who play an enormous amount of great jazz, blues and pop. Music is subjective: if I don't want to listen to something I'm not necessarily saying it's no good, it just doesn't suit my ears. The beauty of radio today is you can pick and choose what suits you. The world's our oyster now, radio wise.'

A successful musician with the band Catatonia, one of the biggest rock acts of the nineties, and a with a subsequent string of solo albums to her name, Cerys Matthews was awarded an MBE for services to music in 2014 and has been presenting on 6 Music since 2008.

'It just happened by chance,' she says.

'I hadn't thought it through or anything. I bumped into the editor of 6 Music at a Sharon Shannon gig at the Boogaloo in north London and he said, why didn't I come and sit in for Stephen Merchant this particular week. "Really," I thought? It wasn't something I'd ever thought about doing, but, well, why not? In that first show I was playing things like Asha Bhosle next to John Denver, and it was during the run-up to Christmas so I was playing all the tracks that reminded me of Christmas – Louis Armstrong, Ray Charles, Nat King Cole – just things I happened to like. I'm a huge Bob Dylan fan and I loved his *Theme Time Radio Hour*, it absolutely appealed to me because I love old music, the old 78s – you don't hear that kind of stuff very often because the quality of the sound is a challenge – and that was probably my main influence. From the start I had a completely free hand in what I played, and having given me *carte blanche* ten years ago obviously it would be hard for them to say I couldn't play whatever I liked now. I often remind them of the Pandora's Box they opened with me.'

Cerys spent a couple of years deputising for presenters such as Merchant and Marc Riley until she eventually took over the Sunday morning slot in 2010. Sitting next to Jonny during the show I watched as the tweets, texts and emails came in throughout the programme, either comments on a track Cerys had played, suggestions for tracks she might play, birthday dedications or just thoughts on the show in general. A parent messaged to say his four-year-old daughter was enjoying listening to the show for the first time. A man emailed to say he was using the show to help him prepare for his first ever stand-up comedy gig that day, while a woman texted to thank Cerys for the 'golden music' she was playing to accompany her painting her bedroom gold. The comments ranged from what people were cooking for their lunch to an explanation of how the months of the year came to be named in Ukrainian; and they came from locations as diverse as the Isle of Lewis and Uttar Pradesh, an astonishingly multinational audience all brought together by a single radio show and the warmth and enthusiasm of its host.

'When I started I hadn't accounted for the instant reaction online, all the texts and emails – the most beautiful conversation started and, years later, it's not finished yet,' she says.

'The show has varying types of listeners: casual listeners, music fans, there are record labels listening, pluggers, management, bands and musicians. We get messages coming in from all over the world, just today we heard from India, Spain, Ethiopia, Ukraine and New Zealand – and they're all listening at different times of the day. I get people writing in from sheltered accommodation and I get people writing in from birthing rooms in hospitals: both ends of the life cycle, and everyone in between. In the modern world you have families who are spread all over the globe, and some of them are listening to the show. We're the link. It's like they're in different places but looking up at the same moon, making that triangle, only with us the tip of that triangle is radio. No other format could do that. The romantic image of radio has always been of families gathering around the radio set and tuning in to the world, and it has that power. It's an innocent tool of communication in its own way because it's just sound – you have to use your inner imagination a lot – and that's a comfort in itself.

'It's just brilliant, this huge network of people who love music, all kinds of music, and they're all here every Sunday. It's a particularly sweet day to do radio because for most people Sunday is down time: you don't have to go to work and you can potter about in your pyjamas with the radio on. People are messaging in saying they're listening to me while cooking the Sunday roast, doing a bit of decorating, or putting together something from Ikea. You get these insights into people's lives in a way I don't think you get with television. When I'm out and about I meet people who tell me they always listen on a Sunday – "you're in our kitchen, you're in our lounge". You wouldn't say the same thing to Dimbleby about *Question Time* because he's not in your lounge, you're in his. TV and radio are completely different from each other in that way. You have to concentrate on television, stop whatever you're doing and look at it, whereas radio accompanies you through your day. You don't have to stop your life for radio because the two become entwined; radio becomes a part of your world, whereas with television or film life stops and you enter *their* world.

'The nature of radio is to be instant and spontaneous. There's just a tiny core team putting this show together. I've done a few bits of TV and you find so many different people on the team, you have shots from different angles using different lenses, and it's time-consuming and expensive – all for a single minute of film. We just did three hours of live broadcasting and it was just me, Angela the producer and Jonny.'

For someone clearly so passionate about the medium of radio Cerys grew up feeling ill-served by it.

'My childhood was during the eighties when Mike Read was doing the Radio 1 breakfast show and playing stuff like the Del Fuegos' "I Still Want You". I remember listening with the cassette ready, pressing "play" and "record" and ending up with all these mix tapes with that hiccupping noise at the beginning and end of each song where you weren't quick enough with the buttons. That said, I was very particular about music from an early age, and the eighties weren't really my cup of tea. I was more interested in finding old recordings. I've always been interested in all genres and all ages of music, hence the show I do now. I spend most of my life collecting music anyway so I've got a huge reservoir that I feed into the show, and with much of it the influence of radio is obvious. I love pre-revolutionary Iranian pop music, for example. I've got sitar versions of Beatles hits from the sixties, which all came about through the artists hearing things on the radio. Through radio doors have opened and connections have been made for generations now. I've interviewed Baaba Maal, who told me about growing up living in a little village in Senegal listening to Hendrix. Selda, who I interviewed, is from Turkey, where she heard Connie Francis on the radio who became her biggest influence. The world is a small place when you have a radio.

'I spend a good day and a half preparing the show and try to theme it depending on what's going on. This month we're looking at festivities and traditions tied in with the arrival of spring: there's always something on which to hook the programme, otherwise there's just too much choice. There's still room for spontaneity, though. I programme the show at home, selecting the music and spoken-word stuff and turning them into digital files to play on the show, but you never know what's going to come in on the day. Sometimes a show can present its own theme somehow, which I find fascinating, and it's one of the reasons I love doing live radio. Sometimes, very rarely, circumstances will dictate that I have to pre-record, and that's not the same at all because I'm just talking and not thinking. That can lead to some funny old links, and you lose a lot of spontaneity.'

Jonny had handed me a copy of the running order as the show began, a wodge of paper packed with the tracklisting and some background to each song, but other than hitting the news on time at half-past the hours and running a couple of trails every hour the programme retained a fluidity and spontaneity that weaved through the running order. To balance the kind of structure necessary for a live show that packs in an eye-watering amount of music and speech with keeping it fresh and spontaneous and open to digressions, tangents and diversions is extraordinarily difficult, but the team of three – plus a studio manager looking after The Fernweh in the live room – made it look and sound effortless.

'Part of the joy of this show is crafting what tracks you put next to each other,' says Cerys.

'In the past we've had things like talking whales and squeaking frogs. The best one was the sound of mating haddock mixed with the *Dr Who* theme, while next week we'll have a snoring hummingbird. All sound is music. It's only our definitions, the finessing we do to it, that change nature's sounds into what we call music. It's all the same spectrum of sound, though, and to me if there's a context behind it then it's worth playing. I've noticed that now more than ever musicians are feeding more and more natural sounds into songs, which is curious considering the increasingly sedentary and indoor lives we live, yet there's been a huge increase in songs featuring blackbirds, the sea, bees, all these sounds being fed into music. I also like pulling out archive recordings of poets reading their own stuff, like Ben Bach today, I just threw him in. I think it's like stitching a quilt: while your focus may be quite narrow you still have to keep in mind the bigger picture, the journey of sound you're constructing, otherwise it would be a bit uncomfortable to go slamming from, say, Mos Def into Ben Bach into Julie Murphy.

'It's quite nice being able to hear the different keys of the songs too, so I often match keys in places where I think we need a bit of sense and comfort. I love that. I like editing songs as well – sometimes you can hear parts of songs that go better with parts of other songs, so I'll take whole intros out, for example, which is a lot of fun. Or I'll experiment with overlapping, which is what I did with the mating haddock and *Dr Who*. But for all the day and a half of preparation I do there's so much information coming in during the show – the tweets, emails, texts and studio guests – and knowing you want to cover so much it's skin of the teeth stuff. Luckily I like doing things by the skin of my teeth.'

As well as playing music from around the world Cerys has, over the years, invited a wealth of guests onto the programme who might not necessarily be heard interviewed anywhere else. When we meet she's been trying to confirm Helen Sharman, the first British astronaut, for the show, and has also had soil experts and scientists in the studio, anyone who might have a fascinating and engaging story to tell.

'I had Allen Toussaint on a few years ago,' she recalls, of one of her favourite interviews,

'...who produced people like Lee Dorsey and the Meters as well as being a great musician in his own right. When I have musical guests on I try to second guess music they'll love and play it as a kind of gift for them, a thank you for coming on. For Allen I lined up some Snooks Eaglin, known as the "little Ray Charles", a New Orleans native and one of my favourite artists. As soon as the music started Allen said, "Ahh, Snooks!" I said, "Yeah, do you know him?" And he said, "Yeah, I used to be in a band with him in New Orleans when we were 16. He used to drive us around in a car." I said, "Really? Well, when did he go blind, then?" "Oh, he's *always* been blind." So there was this blind teenager driving these other teenagers around New Orleans and they're shouting "left", "right" to him, and that story told me more about Snooks Eaglin and Allen Toussaint than pretty much anything else.

'Another one I remember as a highlight was Terry Cryer, this amazing photographer from Leeds who took pictures of all the jazz greats: Coleman Hawkins, Louis Armstrong, Sonny Stitt, all the big jazz stars. Sister Rosetta Tharpe was maid of honour at his wedding and I love having people on like that, people who've been around and who'll talk about the characters they've met through the stories they have. Terry told me all about how he became a photographer by knocking up a fake pass to get him into gigs because he wasn't a professional photographer, he was in the army. He was a real jazz guy, a cat, from the fifties, and we were live on the radio and he was quoting Sister Rosetta, listening to this singer, and out came this inappropriate word. Live on air. He didn't mean it maliciously, of course, it was a story of an African-American singer giving a compliment to a white singer, it had come from a place of love and respect, but at that moment I was looking through the glass expecting to see my P45 being delivered. People had heard my intake of breath and the whole system was flooded with people saying not to worry about it and how much they loved Terry. I apologised if anyone was offended and we moved on. Terry was a lovely man, I'll always treasure his company.

'I also loved having Tom Paley on the show, someone who'd toured with Woody Guthrie, played with Leadbelly, taught Ry Cooder the guitar and kickstarted the folk revival in Greenwich Village. Talking to people like Tom you're engaging with this person who is one connection away from legends, people of mythical stature, and that to me is the joy of what I do. It's not just playing music to make connections in the present it's also forging a connection with the past. It's a huge privilege to be able to talk to them and get them to share their stories with us all. To me that's the ultimate.'

Cerys Matthews is a busy woman. When we speak she's about to take over Radio 2's Monday night rhythm and blues show from long-time presenter Paul Jones and is just putting the finishing touches to a book she describes as 'a food version of the radio show'. She also runs a music festival, not to mention bringing up two children with her husband. Given that most of her spare time seems to be spent hunting out new music it's a wonder she has time to listen to the radio at all.

'Oh, I love Radio 4,' she says. 'I've always listened to Radio 4 even as a child and as a teenager. I'm not so keen on the comedy or drama but I love the news, the documentaries, *PM*, *Broadcasting House* and *Front Row*. I also like John Kelly on RTÉ in Ireland, he's great, plays a brilliant selection of tunes that are hand-curated by a music fanatic and he's a voice you can trust. When I lived in America I loved NPR and WWOZ, and when I'd go on long drives, crossing from state to state, I loved picking up different local stations, trying to dial into something I'd never heard before. It was full of ads, of course, and that's one reason to be grateful for the BBC, because it's advert-free and adverts are excruciating.'

When I follow this up by asking what she would change if she was put in charge of British radio her reaction is so anguished that I almost feel the need to point out that I don't actually have the authority to bestow that kind of post.

'Oh, oh wow,' Cerys gasps, and goes quiet for a moment, chewing her lip. 'Oh,' she says again. 'Let's see. Well, put in charge I would change Radio 4 by taking drama and comedy and putting them onto a different channel, and I'd probably introduce a bit of music, some old 78s, just a little bit here and there, just a minute tweak. I mean, it's a big change for Radio 4, and there'd be letters . . .'

She goes quiet again.

'But you know what?' Cerys says. 'That's kind of a god role. I'd feel uncomfortable with that. In general I would urge us not to dumb down on the radio. Great journalism and the idea you can trust and find truth is really important for the future. People like Jenni Murray, Mark Mardell, Andrew Marr . . . these are people you learn to trust and get to know their way of thinking.

'I can tell you what I'd avoid, that's a bit easier. I definitely wouldn't go down the American route where you have to have a different channel for everything, because it's so heavily programmed. If you're white you listen to that, if you're black you listen to that, if you're Latino you listen to that. It's so polarised and tribal and I'd hate to see that happening here. If we can avoid that then I think we'll be OK.'

She picks up her two bulging carrier bags of books and CDs and we head downstairs and part company outside Wogan House. As we prepare to head off in different directions Cerys raises a forearm weighed heavily by a Sainsbury's bag full of her homework for the week ahead and shakes my hand.

'People say with AI, algorithms, Spotify and streaming that the power of radio is waning, but I think the opposite is true,' she says. 'You learn to trust these voices and the voices are companions. No AI or algorithm is going to be programmed so finely, picking guests, asking questions and judging their answers for a response. In terms of algorithms I can't see that they're going to able to pull out bits of archive and weave them in from classical to west coast G-funk. It could ferret out similar sounding stuff or similar genres, but when you're weaving together other genres, spoken word, and archive, well, I don't know. I'm hoping I'll keep the job for a few years yet.'

26

'It was Basically Boys Playing with Stuff, but They Also Had a *Vision*': Radio Lessons in Hilversum

All this time I had still been steadily accumulating old radios. I hesitate to use the word 'collecting' as that implies some sort of method and specialist knowledge. I was employing none of the first and certainly possessed none of the second. There was no discernible theme to the kinds of radios I was picking up in junk shops or that would turn up in the post with an eBay slip in the packaging. They were literally just old radios. Empty surfaces around the flat began to disappear beneath transistor and valve sets of various vintages in various states of repair. There were little battery-operated portables and ancient hulking Bakelite things that got hot and filled the room with the smell of burnt decades. There were radios of different colours and shapes made by different companies – Ekco, Roberts, Regentone, Perdio – and I'd buy them purely because something about them caught my eye. Some of the radios worked, some of them didn't. Some were definitely nice pieces of design, others were cracked, discoloured and had bits missing, yet something about each of the transistors had appealed to me. Around once a week the postman was ringing the doorbell with another package and Jude would find me at the kitchen table elbow deep in cardboard, showering the floor with polystyrene chips as I rummaged for grip beneath another ancient wireless. She'd raise an eyebrow, and I'd look sheepish and say, 'OK, that's it now, I have enough old radios.' Every time I said it I really meant it. As weird midlife crises go this Radio Accumulation Syndrome I'd developed was, like most of the radios, pretty low wattage. They were cheap, they were harmless and, unlike a two-seater sports convertible or a leather jacket, didn't make me look a twat in public. The only thing a psychologist might have been able to fuse to an attempt at hauling back a fast-disappearing youth was that the radios were all, without exception, older than me.

After a while I realised they did have something in common, after all. When I'd pick one up in a shop or squint at photographs on eBay I'd find myself scrutinising the waveband display. There'd be a range of place names on different sets, but I came to realise that I'd always make sure Hilversum was one of them.

Brussels, Toulouse, Stavanger, Allouis, Kalundborg or Athlone: if they were there, fine, if they weren't, no bother. But for some reason – and it took a while recognise this particular foible within a foible – I was acquiring old radios that had the name of a small Dutch town on the waveband, the more prominently the better. It was the first name I remember seeing on my dad's old radio and the very sound of the word Hilversum seemed to represent the hiss, hum and mystery of the medium that had facilitated my mental escapes from dull suburbia. For all my foraging in history and sharing my love of the medium with some of its most interesting exponents, could the grail of my search for the magic of radio possibly lie in, of all places, Hilversum?

Twenty minutes south-east by rail from the Amsterdam's Centraal Station lies Hilversum. That morning I'd taken a train to St Pancras then a Eurostar to Brussels. From there I joined a train to Amsterdam that crept across foggy, silent flatlands on a dank and freezing day, changing trains again in the Dutch capital for the short run to Hilversum.

From the impressively sleek brick, steel and glass railway station I walked around the tidy, low-rise town centre. There wasn't much to see, to be honest, beyond the Willem Marinus Dudok-designed modernist town hall from 1931. It's a dramatic building with few windows and assertively sharp corners that appears to mass up towards its bell tower, whose chimes were heard marking the hours on Dutch radio for decades until the 1960s. In a corner of the town hall gardens stands a sculpture of 15 tall, bronze sunflowers looking towards the sky, their leaves interlocked, a memorial to the 15 people from Hilversum, including three whole families, who were killed when the Malaysia Airlines flight MH17 was shot down over Ukraine in 2014 *en route* from Amsterdam to Kuala Lumpur. Their names are engraved on a granite bench facing the sunflowers, which became a symbol of the loss when a journalist who'd been to the crash site presented seeds from the Ukrainian field where part of the wreckage landed to a relative of one of the victims.

In a town the size of Hilversum, whose population is around 90,000, the impact of a tragedy like MH17 resounds through the entire community. Those fifteen people, adults and children, had friends, relatives, work colleagues, fellow pupils, sports teammates – there would be hardly a person in the town not directly affected by the loss of those families, and the elegant, eloquent memorial placed in a quiet spot close to the hub of the municipality is somewhere those people can come to sit, reflect and remember.

The unfolding news of the tragedy would have been broadcast to the nation from a complex north of the centre. Hilversum's Media Park, colloquially known as 'Hillywood', is home to almost all the Netherlands' broadcasting institutions. It's a curious location on the face of it, as if the BBC and all the major commercial broadcasters in Britain were huddled together on an industrial estate on the outskirts of St Albans. And Media Park doesn't present the most appealing vista for one of radio's most romantic names – a scatter of buildings of varying vintage apparently designed specifically to have no architectural sympathy whatsoever with each other, next to a major roundabout. There are clutches of satellite dishes, some of them so enormous you sense they are capable of eavesdropping on some whispered gossip passed around Alpha Centauri B, and so many direction signs and speed limit notices you feel there must have been some kind of sponsored sign-a-thon that raised an absolute fortune.

Despite the park's sprawling nature and bewildering signage I recognised instantly the place I was looking for. I couldn't really miss it because the Netherlands Institute for Sound and Vision is a startling-looking building: a giant box clad all round in multicoloured glass panels of blurry wash. A series of terraced pools leads down from the roadside to the building, lending a contrasting serenity to the frantic activity of the building's exterior. The Institute houses the archives of Dutch broadcasting – one of the most extensive in Europe – and a museum dedicated to its history and development. Inside the floor-to-roof atrium of the reception area a rock band was soundchecking at ear-splitting volume while tables were laid for a corporate event of some kind taking place that evening. Fortunately the woman staffing the reception desk happened to be fluent in arm-waving, pointing and facial gurning in languages other than her own, and I soon found myself in a room on the top floor where Dutch media history specialist Bas Agterberg appeared with a couple of coffees, placed them on the table and closed the door against the racket below.

'Sorry about that,' he said in perfect English. 'The facilities here are available for private hire and we obviously have some kind of event this evening.'

For as long as we could stand it we'd stood on the sixth-floor balcony and looked out over the atrium as Bas pointed to where you could see right down into the several subterranean floors that house the archive.

'We have over a million hours of material and it's being added to all the time,' he shouted over the din. 'When I first came here ten years ago it was 800,000 hours – an enormous variety of films, television and sound. The broadcasters all share the facilities, including the record library that's part of our archive. A copy of every record ever released in the Netherlands is here.'

Between Christmas and New Year's Eve every year the Institute hosts the annual public vote Top 2000 chart rundown, setting up a studio roughly where the band were comprehensively demolishing Oasis's 'Don't Look Back in Anger' for a show that runs 24 hours a day during the downtime between the two end of year festivities, for which half the population of the country tunes in. 'Bohemian Rhapsody' wins almost every year; sometimes 'Hotel California' gets a look in.

Despite being faced across the table by a scruffy bloke over from England on some vague premise regarding the radio history of Hilversum Bas, who looks strikingly like a blond Christopher Eccleston in a t-shirt and fitted blazer, generously gave over an entire afternoon to filling me in, in flawless English, on how Hilversum came to be the centre of Dutch broadcasting, showing me round just a small part of the Institute's incredible archive before finally letting me loose in the museum. First he alerted me to the surprising structure of radio and television in the Netherlands.

'When radio started here in earnest in 1925 we had five public broadcasters representing different parts of society: the liberals, the socialists, the Catholics, the Protestants and the Free Protestants,' he said. 'These all still exist and are all based in Hilversum. These organisations were also awarded licences when television was introduced after the Second World War, a moment that could have made things a little different. There's always been an argument here to have a single public national broadcaster like the BBC, while others claim that it's a good thing to have all these different perspectives. Dutch society is pillarised that way: the football clubs and shops were all established in these five segments of society too. It wasn't completely sectarian, they worked together on many things, but there were always very clear divisions between them.'

We'd barely been talking for five minutes and already I could tell Bas had an encyclopaedic specialist knowledge that was born out of a deep love for the subject. He delivered that knowledge with modesty and humour despite the fact that he was faced with someone who knew less about his subject than probably anyone he'd ever met.

'The five groups had all begun as radio amateurs, enthusiasts who got themselves organised,' he continued. 'It took until the sixties but other organisations with a less ideological grounding can apply for a public broadcasting licence, so there was an expansion in the number of broadcasters in the seventies and eighties, while in the last few years politics have dictated that some of them merge. This means, for example, that the Catholics and the Protestants are now combined into one broadcaster.

'The origins of radio in Hilversum are in a company founded here in 1918 called the *Nederlandsche Seintoestellen Fabriek*, which means the Netherlands Sign Factory, that sold radio equipment among other things. There is a myth that the marketing manager of the company became one of the founders of the liberal radio station and lived here. There was a suggestion that they start broadcasting from Amsterdam, but according to the myth his housekeeper didn't want to move to Amsterdam, which meant his wife didn't want to move either because she didn't want to lose the housekeeper. Hence the station started in Hilversum and the other broadcasters just followed suit. It's a story that I like very much, but the more likely explanation is that the land to build here was much cheaper than in Amsterdam. Once the companies began moving out to this location it became known as Radio City, as television didn't start here in earnest until the fifties. The first building on the site was the one next door, a music building, and then the TV studios began arriving and the whole complex was renamed Media Park in 2000.'

That first building, a few yards away in the shadow of the Institute, is a low-rise construction in an L-shape around a forecourt. It's where, in 2002, the far right politician Pim Fortuyn was assassinated as he walked to his car after giving a radio interview in the run-up to that year's general election. There's a small brass plaque on the spot where he died, inscribed with his name and the date of the assassination.

For all Hilversum's status as the epicentre of Dutch broadcasting the first radio broadcasts in the Netherlands were made from The Hague.

'They were made by a pioneer called Hans Idzerda,' Bas informed me, 'who had a radio manufacturing company and realised that if he played gramophone records and made announcements over the airwaves then people would hear about it and come to him to buy a radio receiver.'

Hans Idzerda was one of the most important radio pioneers, yet he's rarely acknowledged outside the Netherlands. Despite coming from a long line of doctors Idzerda, born in 1885, became fascinated by electronics and qualified as an electrical engineer in Germany in 1913.

Before long he had set up Netherlands Radio Industries in The Hague manufacturing equipment for use in wireless telegraphy, something that made him popular with the military from the very start of the First World War. In 1917 he persuaded the Philips company to help him develop and market a glass valve that was a much more reliable driver of broadcasting reception than the prevalent crystal-set technology – a valve that was named the Ideezet after three letters of his surname.

In February 1919 Idzerda caused a sensation at a high-profile trade fair in Utrecht by broadcasting wirelessly the sound of a musical box from his stand to a point 1,200 metres away, much to the delight of the watching Queen Wilhelmina. In order to carry out the demonstration he was awarded a broadcast licence with the call sign PCGG – which he would later market as 'Phenomenal Concerts Given Generously' – the terms of which he interpreted very loosely, leading to the construction of a powerful transmitter at his business premises in The Hague at the end of 1919. Impressing the Queen was just the start: Idzerda was ready to do some proper broadcasting. He placed an advertisement in the *Nieuwe Rotterdamsche Courant* on 5 November 1919 advising that the following evening between the hours of 8 p.m. and 11 p.m. there would be broadcast a *'Radio Soirée-Musicale'* which radio owners would be able to hear loud and clear if they tuned their sets to 670 metres. Sure enough, the next night at eight o'clock Dutch enthusiasts fired up their wireless sets, found the waveband and heard the fairytale tones of Idzerda's music box heralding the start of a programme of music and speech that would last for three hours.

'I was one of the group of five who gathered in Beukstraat to bear witness that night,' wrote journalist and future bestselling author Herman de Man to Idzerda's widow years later of that night in The Hague. 'How happy and carefree we were back then in 1919. I, the young journalist, delighted with my scoop and your husband, so deeply moved that he was shaking with emotion. Even then we knew we were witnessing a moment of great historic importance.'

It's no wonder Idzerda was wobbly: he'd just put out the world's first scheduled radio entertainment show. Not only that it became a regular event, with Idzerda broadcasting live bands and entertainers as well as playing gramophone records up to four nights a week at his peak. So strong was Idzerda's transmission signal that he even developed a following among enthusiasts in England. In the spring of 1922 so many people wanted to listen to one of the 'Hague concerts' advertised by the Hartlepool Wireless Club dozens had to be turned away because the YMCA hall was full long before the broadcast was due to start. 'One wonders what the people who once worshipped at the old church of St Hilda's but a few yards away would have thought had they seen their successors sitting in Hartlepool hearing music caused by a needle on a revolving disc in Holland, 350 miles away,' marvelled one of the lucky ones who'd gained admission.

So popular did the Hague concerts become that the *Daily Mail*, spying an opportunity similar to their sponsorship of Nellie Melba's Chelmsford broadcast, put up, in the summer of 1922, sponsorship for a weekly broadcast in English. The first of these concerts saw the London-based Australian contralto Lily Payling travel to The Hague to be the show's star. Enthusiasts and clubs across Britain tuned in for the eagerly anticipated show, which opened with Payling warbling 'Land of Hope and Glory' and ended in the spirit of international co-operation with the Dutch national anthem. A packed cinema in Nottingham and a gathering in deckchairs in a park in Ramsgate – where enjoyment didn't seem to be marred by interference in the transmission from the North Foreland lighthouse station – numbered among the listening audience.

One man whose enjoyment was tainted slightly, however, was Hans Idzerda himself: so demanding and diva-esque had Payling been during her visit to his premises he said afterwards that he'd had to lie down for a week to recover.

Alas this proved to be the peak of Idzerda's radio career. Once the BBC started broadcasting in the autumn of 1922 British audiences for The Hague concerts began to decline, and in July 1923 the *Mail* pulled their sponsorship. Shortly afterwards the Philips company transferred their growing manufacturing clout from Idzerda's NRI to the *Nederlandsche Seintoestellen Fabriek*, the Hilversum company responsible for establishing the town as the internationally renowned home of Dutch broadcasting. It could have been The Hague instead, which wouldn't have looked half as good on radio dials.

'Philips hadn't seemed all that interested in radio until the later stages of the war when Idzerda co-operated with them to produce a valve for his radios,' said Bas. 'Then when, in the twenties, radio slowly began to find an audience in America and Britain they realised it might be something worth investing in, took over the NSF factory here in Hilversum and started producing radio sets. Within five years Philips was one of the leading companies in the world selling radios.'

Idzerda, alas, would not share in any of that success, even though much of it was built on his character and know-how. Once the *Mail* sponsorship had dried up and he was abandoned by Philips it wasn't long before he went bankrupt. There were subsequent attempts to re-establish himself, but while he might have been a brilliant radio man Hans Idzerda wasn't such a hot businessman. By the time the Second World War began he was running a guesthouse in Scheveningen. In early November 1944 a V2 rocket landed not far from there and Idzerda, who never lost his engineering curiosity, went to see the wreckage of the jet-propelled flying bomb, only to be told to clear off by Nazi troops. He returned the following day, however, and, according to some sources, was carrying off a piece of wreckage to examine at home when the same soldiers stopped him, arrested him then shot him the following day as a spy, almost a quarter of a century to the day after the magical music-box tones of that first 1919 broadcast.

'Many people make a strong case for Idzerda being the world's first broadcaster, at least as we know broadcasting today,' said Bas. 'He planned it, he put an advert in the newspaper and he didn't just do it once he kept going, regularly broadcasting this programme of him talking between gramophone records. Radio hasn't stopped in the Netherlands since 1919, it's been continuous, and I don't think there's anywhere else that can claim the same thing. Every country has its own claim, of course – Belgium claims a transmission from Brussels to Paris in 1914, for example – but whether he's the first or not Idzerda is a very important person to us indeed.'

They may have been late to spot the true potential of radio broadcasting but Philips were quick to react when they saw the opportunities it afforded. In 1927 they founded PCJJ, a shortwave station designed to broadcast from Eindhoven to the Dutch colonies around the world, effectively a Dutch forerunner of the BBC's World Service. At the end of May that year Queen Wilhelmina, who just eight years earlier had been so taken with Idzerda's trade fair broadcast, was able to address her subjects in both the East and West Indies. A few months later Eddie Startz joined to host the magazine programme *The Happy Station* to commence one of the most extraordinary broadcast careers of any presenter from any nation. The irrepressible Startz presented *The Happy Station* from 1928 until December 1969, a continuous run broken only by the Nazi occupation of the Netherlands; a career presenting the same programme that began before the discovery of penicillin and ended with men in space.

A former seaman from the neutral territory of Moresnet, once a sliver of territory between the Netherlands, Belgium and Prussia, Startz was destined to be an internationalist. He lived in America for a period in the early 1920s, washing dishes, waiting on tables and working for a while as a travelling salesman, before returning to Europe to join Philips just prior to PCJJ first transmissions to the world. Mainly thanks to the fact he spoke a host of languages as a result of his travels Startz was engaged to broadcast on the company's global short-wave station, developing *The Happy Station* as a magazine programme designed to give an insight into life in different parts of the world, fuelled by Startz's boundless energy and irrepressible optimism.

'Because of the time difference many of the shows went out during the night here, so they were recorded onto lacquer discs in the daytime for later broadcast. This means that we're lucky enough to still have many of them in the archive today, a really invaluable resource,' said Bas, before opening his laptop to play me a scratchy disc recording of Startz from 1936, during which he switched seamlessly between Dutch, French, German and English. Then there's a flurry of typing and clicking before a video file loads.

'This is a newsreel from the thirties that marked ten years of the Dutch World Service. It features footage of Eddy Startz in a studio in Hilversum when the service was based in a building in the city,' Bas noted, tapping enthusiastically at the keys. 'I find this really interesting because you're seeing the work in progress, gaining an insight into how things were done back then.'

Not long before coming to Hilversum I'd watched a British film from 1934 called *Death At Broadcasting House*, an ingenious murder mystery set in the early days of the BBC's headquarters written by the Corporation's then head of production Val Gielgud, brother of John, and well worth tracking down for its footage of a Broadcasting House so new the plaster was barely dry. The newsreel Bas showed me looked very similar, featuring lots of earnest young men with oiled hair wearing heavy worsted suits and thick-framed spectacles, carrying pieces of paper around, perpetually in a hurry. The only difference was that nobody in Hilversum was being strangled by a mystery assailant. Or so I assumed. The newsreel celebration showed announcers in different countries namechecking 'Radio Netherlands, Hilversum' in various languages, followed by some Dutch children reading out messages to parents working abroad. Then there was Eddy Startz, who with his sharp features and high forehead could have passed for a member of Kraftwerk if he hadn't insisted on smiling all the time. 'Hello everybody, everywhere,' he chirped, his enthusiasm fizzing from the screen even 80 years later. 'Good morning, good afternoon, good evening! Once more, this is *The Happy Station*. Is everybody happy? Good! Now I'm inviting you for a traditional nice cup of tea, and some happy company!' With that he held up a tea cup to the microphone, in which he vigorously rattled a teaspoon.

'The Radio Netherlands Worldwide service ended in 2012,' lamented Bas. 'It was part of the Department of Foreign Affairs, whose view was that everyone can listen to radio online now so we don't need this service for our own citizens around the world. They didn't like the idea we had to make radio programmes for other countries in order to disseminate our view of the world. It makes sense. I don't think our values differ very much from those of the BBC World Service, for example, so Radio Netherlands didn't distinguish itself enough any more to justify the expense.'

The government cut funding to the global network by a whopping 70 per cent in 2011, resulting in the final broadcasts of the historic service being made in May the following year.

'There is still a Dutch world service but, instead of broadcasting, it trains people from developing countries to make radio,' said Bas. 'We recognise there's a pressing need for media in a democratic society, so there is still some funding available to train people from other countries in order that they can return to improve radio services in their homelands.'

Bas took me down to the depths of the building to show me some of the archives. We descended to the subterranean levels, and he hauled open metal door after metal door, flicking on light switches with an echo that rebounded around the cavernous, windowless breezeblock rooms full of racks of shelving – upon which sit hundreds upon hundreds of old radios, from ancient crystal sets to novelty transistors (there's one of the former US President Jimmy Carter jumping out of a peanut). We entered rooms packed with studio consoles, televisions and cameras. There are drawers full of old headphone sets and little bits of, well, just *stuff*, the use of which I couldn't even begin to guess at. Bas led me to a rack of shelving containing rows of old bound magazines.

'These are radio guides from the very early days,' he said. 'Everyone would subscribe to one: like joining a club, you became a member of whichever of the five pillars you belonged to and you'd receive their radio guide as part of your membership. This went on for decades – right up until the mid-nineties everyone was a member of a broadcasting company. Sometimes I speak to groups of school students who come to the Institute and I ask them if they're a member of one of the broadcasters. Six, maybe seven years ago they might have said no but their parents were, but now they struggle to even understand the basic concept of a broadcaster, that there might be a single source for their broadcast consumption. Most of them don't really watch television or listen to the radio any more, they just get it all online.'

He pulled out a volume from the Catholic broadcaster and flicked through some of the issues. 'I like the fact the programme listings are international,' he said, holding out the huge book of yellowing paper to show me. 'Look, it's not just the details and timings of their own shows, they also give the listings for London, Vienna, Munich, Brussels, Daventry, places all over Europe.'

Listening to the radio in its early days was a truly European experience. It was as normal to tune in to a dance band in Sweden as the Savoy Orpheans in the Strand, as routine to be sitting in one's armchair listening to a Schumann piano concerto from Toulouse as the BBC orchestra. We think nothing today of flitting across the Channel on a whim, the rest of Europe never more accessible, but how often would we sit at home in the evenings and wonder what's on Radio Stavanger, or ask 'What do you fancy tonight? *Coronation Street*? Or some Schubert on the Leipzig programme?' Half of Europe had not long stopped trying to blow up the other half, yet here was radio crossing boundaries and borders and uniting a continent in the ether. That someone in Delft might tune in to Daventry, or a Norwegian pick up the Prague programme: this was the power of radio manifesting itself right from its earliest days as the continent moved from a decade of appalling conflict into a decade of hope and co-operation.

In one of the archive stores Bas proudly indicated some of Idzerda's original transmitting equipment on which he would have made that first broadcast. Among the breezeblocks, concrete pillars and gun-metal grey shelving a hotchpotch of dark wooden boxes and circular antennae are polished and cared for like the antiques they are – items that exude a cultural and historical dignity beneath the stark lighting. In these pieces of equipment there are stories – stories broadcast and stories of broadcasting and pioneering experience – that lend the old pieces a quiet dignity. These items were built, handled and operated by Idzerda himself. Bas squatted down in front of a lower shelf and showed me a glass-fronted wooden case, inside which was a reel of ancient wire.

'That's a level gauge,' he told me. 'It was used during the First World War to detect espionage stations and Zeppelins flying over to

London. It's something Idzerda created for the Ministry of War early in the conflict, and it's probably the reason why he was later awarded the permits for his radio station.'

Then he opened a drawer nearby to reveal a row of the original glass valves Idzerda had developed with Philips in the later part of the war; valves that revolutionised the quality of Dutch wireless enthusiasts' receiving sets. He lifted one out, carefully removed the clear plastic lid of its protective case, and showed me this beautifully engineered and crafted century-old triode valve nestled on a bed of cotton wool like a fossil from another age. It'll never be used again but it looked as good as the day it was made. I tried to imagine the sense of excitement a wireless enthusiast might have felt when he first bought one of these, hurrying home from the electrical shop to his house in, say, Alkmaar, closing the front door, wiping his feet, greeting his wife with a peck on the cheek and going straight to the room where he'd built his wireless set, carefully removing the valve from its box and introducing it to his system. As it slowly warmed up and began to glow it must have looked absolutely magical, its soft light bathing the new owner's face. Then he'd put on his headphones, turn his dials and wait for the world to talk to him like it never had before.

'These amateurs, usually men, usually white men, 40 or older, they loved playing with this stuff,' said Bas with a chuckle. 'Some things never change.'

We stood in front of Idzerda's broadcast equipment and I felt an extraordinary privilege to be there. These items – the lovingly polished boxes with their hand-painted lettering over the switches and dials, the pieces of metal studded with nuts and bolts that were tightened by Idzerda himself – are priceless relics of broadcasting history. All the radios in the vaults, all the ones I'd been gathering at home, they all had the word Hilversum on them because of work started by Hans Idzerda using the items on the shelves in front of me. The awestruck crowd in the Hartlepool YMCA, the people snoozing in Ramsgate deckchairs as an Australian diva serenaded them from across the North Sea – that voice had come through the equipment in front of me, some of which was constructed from something as simple as a cigar box.

'What I find interesting about these people is that they were like the internet pioneers,' said Bas. 'It was basically boys playing with stuff, but they also had a *vision*, a vision that they carried as far as they could and changed the world.' Bas paused for a moment, genuinely awed even though he'd stood in front of Idzerda's equipment hundreds of times. 'The sheer imagination that people had,' he said. 'What's fascinating is that while I'm proudly telling you about Dutch pioneers like Idzerda the same thing was happening at the same time in Britain, Germany and many other countries. It wasn't the case that in, say, 1910 we did something then a few years later the Germans found out about it and advanced the technology and so on. Instead, all over Europe and the world there were people working away by themselves, creating, having that vision of broadcasting and having it all at the same time.'

Bas left me to have a look around the museum, an impressive place, dark and cavernous with sections on different aspects of Dutch television and radio, albeit most of it lost on me. You can have a go at DJing in a mocked-up radio studio, using jingle cartridges and cuing up records, and even be filmed presenting a television news bulletin. On the second floor there's a display made up of two giant waveband displays, a mock-up of a massive radio featuring all the place names of the dial, one side facing the other at head height, and you pass along as if you're the needle and someone's turning the dial. I walked slowly along the radio cartography of Europe – Roma, Jaarsveld, München, Beograd, Motala, Zeesen, Daventry, Lodz, Oslo, Koblenz, Malmö, Helsinki, Lyon, Riga, Bucureşti, Hamburg, Strasbourg – and in my head I could hear the static and the bursts of passing stations as I moved slowly along the dial.

The sun was setting by the time I headed back to the station – all broad, vivid purple and orange stripes – and I walked around thinking about how I'd come to Hilversum in search of something but was still not sure what it was. The answer, I'd learn, was much closer to home.

27

Listening Out at Rampisham Down

The towers fall one by one and without ceremony. Men in luminous jackets and hard hats high-step through the long grass with dew-darkened toecaps. A drop to one knee, a spray of bright orange sparks and, almost imperceptibly at first, the tip of another latticed metal giant begins to glide towards the ground with the same melancholy dignity as the rest. A graceful arc accelerates until there's a metallic *crump* that's whipped away across the hills by the wind.

Thirty-four of these towers stood at the highest point on Rampisham Down in rural west Dorset. Seven hundred feet up on the hill, some of them 350-feet high, this was a vantage point from where the BBC World Service was once catapulted to the farthest outposts of the globe. The masts were a comforting presence on the horizon, almost protective. As long as the towers were there – implacable, authoritative, with the spars that emerged from just below their peaks making it look like they were keeping watch in all directions – all was right with the world. There's barely anyone alive now who would remember a time before the Rampisham transmitters, and as the towers started to fall the spaces appearing in the skyline looked more incongruous than the clutch of steel goliaths ever did among the undulating hills of Hardy country.

When the British Foreign and Commonwealth Office decided to pull its funding from the World Service in 2009 the Dorset transmitting station was one of the first victims of the cuts. The last World Service transmission sent out via Rampisham was on 29 October 2011, broadcast to North Africa by the BBC's Arabic Service on 5790 and 11680 kilohertz, while the last transmission of all followed immediately afterwards, a Deutsche Welle programme beamed to Europe on 6075 kilohertz that ended at 21.59. At 22.00 the station fell silent for the last time and the towers became mute sentinels beneath a canopy of stars.

My parents lived in the shadow of Rampisham Down for 20 years. I'd see the towers every time we left their village – glimpses from the car, or a handy orientation landmark on walks. When I left the city to visit and get away from the noise and hullabaloo I'd be reassured by the towers on the hill. The world and its troubles felt a long way off in time-stilled rural Dorset. All those voices in the air – the news, the discussions, the dramas, the jabbering and gabbling, making a point, avoiding a question, laughing, doubting, singing, reciting, asserting, arguing – all of it swirled around me in the silence.

Whenever I went into Bush House, the old World Service building on the Aldwych in London, and spoke into a microphone, I'd always think of the Rampisham towers, how my half-baked ramblings were being fielded there and twanged around the globe to places I could barely imagine, swirling around my parents' house, whooshing past the windows as they set off on their global odyssey.

Rampisham Down transmitting station had been a feature of the west Dorset landscape since the BBC purchased the 189-acre site at the beginning of the Second World War. The station, with four Marconi transmitters contained in two large halls, was given the unimaginative name 'Overseas Extension 3' and came into operation on 16 February 1941, immediately attracting the attention of the Luftwaffe. Bombs fell close by but the transmitters were vigorously defended by the RAF, with locals telling stories for decades afterwards of dogfights over the villages and fields. Bombs never fell any closer than the other side of the road, and there was some damage to a diesel tank once, but Overseas Extension 3 kept going through it all.

A small, basic studio was incorporated into the bowels of the station, used occasionally to broadcast reports from war correspondents recorded onto gramophone records, and to send coded messages to Resistance fighters across Europe. During the Cold War when the Russians attempted to jam radio signals from western Europe the BBC responded by opening up all their transmitters, including Rampisham, to bombard the Soviet Union with an armada of World Service broadcasts on a range of frequencies – a squadron of sound impossible to deter.

When I used to walk up close to the quiet, grassy field in an out-of-the-way part of an out-of-the-way county it seemed an unlikely combatant on the front line of global affairs. Those towers had secrets, and I wondered if ghosts whispered through them in the night.

The first of them fell in 2013, just one or two initially, before the complex was declared a Site of Specific Scientific Interest by Natural England in August of that year due to the expanse of lowland acid grassland that covered most of the site and had been almost completely undisturbed since the land was acquired by the BBC in 1939. In 2015 British Solar Renewables applied to use the site to install solar panels, eventually locating the solar farm nearby instead but, as part of the planning permission, agreeing to help Natural England bring down the rest of the towers.

There are only two left now, spaced far apart, the spars protruding from the top of the towers facing each other across an empty expanse as if hoping one might tell the other where the rest of them have gone. When I got back from Hilversum I went to have a look – the first time I'd been to the site since my parents moved away from the area a few years ago. Metal fencing surrounded the site and vast, empty cable reels were jumbled against a wall of the vacant office block. I stood by a gate that had an old, dirty sign hanging loosely from the top bar. 'Danger of death' it warned, beneath a yellow triangle on which a stick man was being struck by lightning. 'Keep away' it added, under a big white exclamation mark in a blue circle. The only sound was the wind hissing through the tufty grass. Overseas Extension 3 was once a busy place, thrumming with activity. It had a social club for employees and a range of sports teams, from cricket to chess. Now there were just two lonely towers, a few old buildings and silence.

A commotion of flapping came from deep among the reels, and a buzzard emerged, heaving itself slowly into the air. It cleared the roofline, latched onto the breeze and transformed from a weary, clumsy-looking creature that could barely haul itself off the ground into a graceful monarch of the skies, wings fully extended, gliding and swooping around the nearest tower until it settled on the spar and folded its wings together. Once upon a time that bird would have ridden thermals of voices and winds of words; soaring over a song from Suriname or a weather forecast for the Western Sahara, but now it's just the plain old wind that barrels across the Dorset hills.

Sometimes while visiting my parents I'd lie awake at night with the curtains open, looking up at stars that are invisible in the city, thinking about the voices being channelled through the air just over the hill. The last night I stayed in the house, before my father's advancing Alzheimer's forced a move to somewhere a little less remote, I reached for the little Roberts travel radio I carried everywhere in its wraparound case, clicked it on and pushed it under my pillow, letting the voices leaving Rampisham talk quietly into my ear until I fell asleep.

The buzzard took off again, swooped down, extended its wings and flew off with long, slow, lazy movements in the direction of the village where my parents used to live.

I was beginning to realise what a strong connection radio has with a sense of home. Most of our formative radio listening is done at home: first the transistor in the kitchen as you get ready for school, and the *Archers* omnibus on a Sunday morning as the lunch cooks. Then your first radio of your own, the little one under the pillow, the shifting of your head until your ear is right over the speaker. Then, if you're lucky, your own music system on which for the first time in your listening life radio doesn't sound thin and reedy. The next step is your own home and your own radio for the kitchen. A clock radio for the bedside. A waterproof radio for the shower. Measuring out your life in a succession of radios.

Right from early childhood the radio is party to your most intimate life in a way that television isn't. Your television set stays in one place – you have to go to the television – but you make sure radio follows you around the home.

Up on Rampisham Down watching that bird ease itself from the spar and glide off towards the village I finally understood why I'd embarked on these journeys into radio and, most importantly, why I'd begun accumulating those old valve sets and transistors.

When radio began to take hold during the early 1920s its possibilities seemed endless. It was a time when the nation was still coming to terms with the trauma of the First World War. Three quarters of a million combatants, almost all of them men, almost all young, crossed the English Channel and never came back. There was a huge void in the population – death on a massive scale, each of them a raw, agonising personal tragedy for mothers, fathers, friends, siblings, spouses and sweethearts. There was roughly one death in the conflict for every 60 people in the country. That's three or four on an average street in an average town, each of them mourned by an extended network of people who loved them.

Perhaps unsurprisingly in the years that followed the war spiritualism flourished. So much shock, so much grief, so many unanswered questions, it's no wonder people hoped and believed they could contact their loved ones in the afterlife and mend the fractures in their psyche.

And then there was the wireless. A machine that captured invisible voices from the ether and made them audible. If the voices of the living could be heard in such a fashion, surely the same must be true of the voices of the dead?

Local newspapers from the early 1920s are full of accounts of meetings held in halls up and down the country in which mediums would attempt to contact the dead and spiritualism was discussed in scientific terms. Wireless invigorated these debates as an example of how we'd really no idea of what manner of voices were calling to us through the atmosphere. As late as 1930 when Mars passed considerably closer to Earth than usual wireless enthusiasts and mediums across the country claimed to have picked up tangible messages sent from the red planet. In 1922 a well-attended meeting of the Sheffield Rotary Club agreed a broad consensus that the next step forward from broadcast radio would be a device on which we could tune in and listen to people's thoughts.

Sir Oliver Lodge, a British physicist and one of the key figures in the early development of wireless – often referred to as Britain's father of radio – lost his son Raymond in 1915 when a shrapnel shell exploded over the communication trench in which he was sheltering just outside Ypres. Lodge was a firm believer in spiritualism, through which he'd become a friend of fellow believer Sir Arthur Conan Doyle, and after the war spent long hours in séances with a well-known medium named Gladys Osborne Leonard through whom he was convinced he had communicated directly with Raymond. A year after his son's death in Flanders Lodge published a book, *Raymond, or Life and Death*, in which he set out his scientific reasoning for accepting Leonard's channelling of Raymond from the afterlife as genuine and logged in minute detail the conversations with his son conducted through the medium.

Not long before I'd travelled to Rampisham I'd read a poem by the radio historian and poet Seán Street called 'Wireless', a beautiful meditation on Lodge and his grief. Two of the stanzas read:

I ask my radio to listen for you,
catch your words in the ectoplasm of white noise.
Voices come flooding past with the idea,
But not the sound I've carried like a jewel

within me. In truth each dial's turn I make
is a prayer for you, possibilities
refining the search across silhouetted
horizons of sound. The speaker gapes static.

The day I visited Rampisham was the first anniversary of my father's death. He's buried in the churchyard almost within view of the towers. I had just come straight from visiting his grave.

I'd watched helplessly for four years as this gentle, brilliant, quietly heroic man lost to Alzheimer's disease his eloquence, his bearings, his self-confidence, his motor skills, his memory, his ability to feed himself, his dignity, his ability to communicate. Towards the end of his life I'd visit him in the care home where he spent his last two years, see the absence in his eyes of almost any trace of the father I loved and listen to the stream of gibberish that poured from him in the handful of hours a day he spent awake. I'd listen desperately closely for something that might make sense, something that would indicate he was still in there somewhere. I'd try to lead him in conversation about things we'd done together, even though I knew he had absolutely no idea who I was, as if I was turning the dial through the static hoping I might catch a voice I could recognise. Sometimes I'd make out a couple of words that sounded like him and I'd think, 'Was that . . .?' but then the hint of a signal would be lost and the static of gibberish would start again.

At his funeral we played 'Star Eyes' by Sarah Vaughan as his coffin was brought into the little village church on a small Dorset hill. Dad's first job after leaving school had been at a London office where DJ Pete Murray's wife also worked. Eventually he plucked up the courage to ask her if Pete would dedicate a song to his friends on his Radio Luxembourg programme. Sure enough, 'Star Eyes' by Sarah Vaughan went out dedicated to Dad's gang of mates, beamed across northern Europe all the way from Luxembourg. Sixty years later it was the soundtrack to his final journey.

As the bird shrank to a speck over the village where Dad lies I thought about all the old radios I'd acquired, the heavy old Bakelite valve ones and the beautifully designed transistors with all those mysterious place names on the dials, and I finally realised why.

Occasionally I'd pause by one as I passed and drop to my haunches in front of it. I'd switch it on, waiting that few seconds of warm-up silence if it was a valve radio, or getting the immediate hiss of static from a transistor. I'd turn the dial the length of each waveband, one end to the other, and pass along the bursts of speech and music. I'd thought I was taking a brief moment out of my day to marvel at the everlasting magic of radio, but that's not what I was doing at all.

I was listening out for him. Wherever he was I wanted to hear his voice, I wanted to know he was himself again. I wanted to know he was all right.

I also realised that whenever a new radio would arrive and I'd unpack it, have a quick fiddle with it to get it working and place it somewhere in the flat, I wasn't doing it randomly, as I'd first believed. The penny dropped that the radios were situated so I'd always be in sight of one. Whichever room I was in, wherever I was, there'd be a radio. I was making sure I was always in earshot should he ever come through the static.

I climbed back into the car and pointed it east, the milky yellow wintry sun low behind me. As I pulled on to the A35 I remembered a Christmas Eve when I'd been very late getting away in the evening, driving the other way down the same road on an absolutely clear, starry night. I was the only person on the road in either direction, everyone else was where they were supposed to be, and when I switched on Radio 4 the classic, definitive BBC recording of Dylan Thomas's *Under Milk Wood* was about to commence. There was a few seconds of silence after the announcement and then Richard Burton's velvet urgency: '*To begin at the beginning . . .*'

It was probably the most perfect radio moment of my life.

I thought of my mum, in her new house now, where she'd be halfway up a stepladder painting a wall, making the place her own, and nearby the Roberts New Classic 993 I'd bought her recently, a portable radio she could carry round with her from room to room, filling each with the sound of familiarity as the last echoes of the previous occupants dispersed.

And I thought of my dad, hoping he was out there somewhere, well along the way on his last journey. I'll keep turning those old dials through radio geography, listening hard, trying to pick him out of the ether somewhere among Brussels, Lille, Bucharest, Athlone, Kalundborg, Toulouse, hoping to catch a word or two, hoping to hear about his ride on the last train to Hilversum.

FURTHER READING

Searching for the magic inherent in a listening medium involved a substantial amount of reading. I'm not going to list all the written works I consulted while putting together this book – mainly because I'm disorganised and can't remember half of them – but I would like to mention a few sources that I found particularly valuable.

If the study and appreciation of radio, its culture and history has a national treasure then it is Seán Street. In terms of research I found myself consulting both his *A Concise History of British Radio 1922–2002* and *Historical Dictionary of British Radio* more than any other sources. In addition his *The Poetry of Radio: the Colour of Sound* and *The Memory of Sound: Preserving the Sonic Past* opened my eyes and ears to new ways of listening to and thinking about radio.

I first came across Seán's work through his poetry, specifically 'Shipping Forecast, Donegal' from his outstanding collection *Radio and Other Poems*. Seán's verse captures beautifully both the intimacy and magic of radio. His collections *Cello* and *Talk, Radio: Poems of Transmission* (from which the lines quoted from 'Wireless' on page 318 appear with Seán's kind permission) are also favourites of mine, as is the terrific anthology of radio-related verse he compiled, *Radio Waves: Poems Celebrating the Wireless*.

Paul Donovan's *The Radio Companion* proved to be an invaluable writer's companion, while Asa Briggs' five-volume *The History of Broadcasting in the United Kingdom* remains a timeless monolith of research and erudition. I also enjoyed very much Charlotte Higgins' *This New Noise: the Extraordinary Birth and Troubled Life of the BBC*, while the spirit of the early days at Writtle was captured wonderfully by Tim Wander in *2MT Writtle: the Birth of British Broadcasting*. Jonathan Hill's *Radio! Radio!* is an absorbing account of the story of radio as well as being a terrific guide to vintage radio sets, if you're into that sort of thing. I found my memories of listening to Laser 558 in equal parts honed, confirmed and corrected by Paul Alexander Rusling's *Laser Radio Programming: 10 Million Listeners Can't Be Wrong*, while Kate Murphy's *Behind The Wireless: a History of Early Women at the BBC* is an outstanding account of an important and overlooked story.

If some of the biographical sketches in the preceding pages have fired a desire to know more about some individual broadcast lives, I can heartily recommend *The Expense Of Glory: A Life Of John Reith* by Ian McIntyre, *Marconi: the Man who Networked the World* by Marc Raboy, *And the World Listened: the Biography of Captain Leonard F. Plugge* by Keith Wallis and the autobiographies *Broadcasting a Life* by the extraordinary Olive Shapley and *This–Is London . . .* by Stuart Hibberd.

Online sources in which I can – and do – lose myself for hours include invaluable history sites www.transdiffusion.org, www.offshoreradio.co.uk and Andy Walmsley's outstanding Random Radio Jottings blog at www.andywalmsley.blogspot.com. To hear the recording of Peter Eckersley's recreation of the glory days of 2MT go to www.emmatoc.com.

Obviously the most exciting piece of radio-related writing, however, is when you're a kid and you receive a postcard praising your Jumbly picture that comes from the radio itself.

ACKNOWLEDGEMENTS

It's always an honour to publish a book with Bloomsbury and I'm mightily grateful to everyone there for all their invaluable input in turning a great big wobbling mountain of words into the book you now hold in your hands. To Charlotte Croft, Allie Collins, Ian Preece and Holly Jarrald, James Watson and Jude Drake a great big smack-eroo on the cheek, a curt nod or a hearty high-five, whichever they prefer.

I'm immensely grateful to Charlotte Atyeo for her friendship, relentless enthusiasm for and passionate advocacy of *Last Train to Hilversum* from its earliest stages.

Customary but never less than glittering thanks are due to the world's greatest agent Lizzy Kremer and her inestimable colleagues Harriet Moore and Maddalena Cavaciuti at David Higham Associates.

Simpering gratitude goes out to all those who submitted themselves to interview in the preceding pages without whom, well, you'd have just been stuck with me droning on through the whole thing, for one thing. Thank you Corrie, Dotun, Cerys, Arthur, Sandee, Alex, Charlotte, Bas and Jessie for your time and insights.

David Prest, head honcho of Britain's finest independent radio production company Whistledown Productions, has been a good friend and constant source of generous advice and encouragement to me on and off the airwaves for many years, but I don't think I've ever said thanks. So . . . thanks.

I'm also grateful to the following splendid folk for their assistance, company, advice, opinions, memories, eye-rolling at yet another obscure radio anecdote, and strong drink when needed – in some cases a combination of all of these. In no particular order: Thomas de Bruin, Liz Buckley, Dale Shaw, Richard Courtice, Tom Teasdale, Kevin Dawson, Harry Graham, Hannah Bushell, Constance Ratazzi,

Gary Bales, Angela Clarke, Miles Hunt, Bernard Sumner, Seán Street, Steve Morgan, Dan Berlinka, Seb Emina, Karina Buckley, Sheryl Garratt, Mark McGuire, Peter Finch, Lorcán Leavy, Abigail Rieley, Jasper Copping, Sinéad Gleeson, James Pennock, Chris Skudder, Claire Deakin, David Parsons, Marie Parsons, Liz Campbell, Erica Nockalls, Catherine Rotte Murray, Ian Darby, James Brown, Dan Donnelly, Eero Hämeenniemi, Colm Tobin, Roy Clare, Eleanor Tiernan, Daisy Buchanan, Michael Grant, Mike Gerrard, Maura Foley, Allison Burke, Lee Watt, Rosita Boland, Samira Ahmed, Jim Kimberley, Isobel Kimberley, Paddy Courtney, Matt Kelly, Perry Bartlett, A.J. Lampe, Nadine O'Hare, Barrie Birch, Sam Bryant, Nigel Bryant, Mark Warren, Melissa Shales, Penelope Shales Slyne, Richard Thomas, Alan Townsend, Fritz Hermann, Alex Ratcliffe, Kate Morrish, Tom Muckian, Margaret Blake and Maria Boyle.

As ever, the largest chunk of gratitude goes to Jude, without whom, well, I'd have starved long before the deadline for a start. No one could be a bigger, better or more relentlessly cheerful and loving support.

This being a radio book I should probably finish in the traditional manner by saying hello to anyone else who knows me.

INDEX